One Hundred and Sixty Minutes

The Race to Save the RMS Titanic

William Hazelgrove

Prometheus Books

Guilford, Connecticut

Prometheus Books

An imprint of Globe Pequot, the trade division of
The Rowman & Littlefield Publishing Group, Inc.
4501 Forbes Blvd., Ste. 200
Lanham, MD 20706
PrometheusBooks.com

Distributed by NATIONAL BOOK NETWORK

British Library Cataloguing in Publication Information Available

Library of Congress Cataloging-in-Publication Data

Names: Hazelgrove, William, 1959– author.
Title: One hundred and sixty minutes : the race to save the RMS Titanic /
    William Hazelgrove.
Description: Lanham, Maryland : Prometheus, 2021. | Includes
    bibliographical references and index. | Summary: "One hundred and sixty
    minutes. That is all the time rescuers would have before the largest
    ship in the world slipped beneath the icy Atlantic. There was amazing
    heroism and astounding incompetence against the backdrop of the most
    advanced ship in history sinking by inches with luminaries from
    throughout the world. It is a story of a network of wireless operators
    on land and sea who desperately sent messages back and forth across the
    dark, frozen North Atlantic to mount a rescue mission. More than
    twenty-eight ships would be involved in the rescue of Titanic survivors,
    along with four different countries. This is a detailed and
    all-encompassing look at the Titanic rescue mission"—Provided by
    publisher.
Identifiers: LCCN 2021005884 (print) | LCCN 2021005885 (ebook) | ISBN
    9781633886971 (cloth) | ISBN 9781633886988 (epub)
Subjects: LCSH: Titanic (Steamship) | Search and rescue operations—North
    Atlantic Ocean—History—20th century. | Shipwrecks—North Atlantic
    Ocean—History—20th
Classification: LCC G530.T6 H39 2021 (print) | LCC G530.T6 (ebook) | DDC
    910.9163/4—dc23
LC record available at https://lccn.loc.gov/2021005884
LC ebook record available at https://lccn.loc.gov/2021005885

Once again,
for Careen, Callie, Clay, and Kitty

Putting about and heading for you.
—12:32 A.M., APRIL 15, 1912, RMS
CARPATHIA TO RMS *TITANIC*

We are putting passengers off in small boats. Women and children in boats. Cannot last much longer. Losing power . . .
—ONE OF THE LAST MESSAGES SENT
BY JACK PHILLIPS FROM THE *TITANIC*

. . . --- . . .
—SOS SENT BY *TITANIC*, ONE OF THE FIRST SHIPS
TO USE THE NEW DISTRESS SIGNAL

You have done your duty. . . . It is now every man for himself.
—CAPTAIN SMITH TO WIRELESS OPERATORS
HAROLD BRIDE AND JACK PHILLIPS

Contents

ACKNOWLEDGMENTS

Thanks once again to my agent, Leticia Gomez, for placing this book in the right hands. And to Jonathan Kurtz and the folks at Prometheus who worked during a pandemic to bring out a book about another catastrophe that to people in 1912 was just as shocking and devastating. Travel was restricted during the research of this book so thanks to the librarians, research assistants, and others who endured the emails and strange calls from a man asking arcane questions about a ship and how to send a wireless telegraphic note. And once again to my family for enduring the author who slips away to a room over a garage to dive into history day after day after day.

Titanic Mythology

On July 29, 1912, Alfred Marsh from Plainfield, New Jersey, was out fishing for bluefish with friends when he saw a brown glass bottle float up behind the gunwale of his boat. According to the *Plainfield Courier News*, Alfred fished out the whiskey flask with his net and noticed some white paper inside. He uncorked the bottle and read the note, scrawled in pencil: "The Titanic is sinking. Goodbye. John James. Fergmaun Road, Cornwell England."[1] The bottle was covered in slime, but the note was dry. Obviously, John James, seeing the ship was going down, threw this bottle off into eternity. A great story. The only problem was, there was no John James on the passenger list of the *Titanic*. John James Ware was listed in second class, and a John James Borebank was in first class. Neither man was from Cornwall, England, which the author of the note misspelled with an "e." Still, the mythology of the last note from the *Titanic* in the bottle persists, and the questions are neatly disposed of.

Forty years later, a young thirtysomething copywriter named Walter Lord put an ad in a local newspaper requesting to speak to any *Titanic* survivors for a book he was writing. Many survivors from the famed wreck were getting on in years, however Lord managed to get a glimpse of that night in April before they slipped away. The last survivor of the *Titanic*, Millvina Dean, died in 2009 at age ninety-seven. The book *A Night to Remember* came out in 1955 and became the bible of the *Titanic*. Up until then there had been a slow-building interest in the fate of the "unsinkable ship" that slipped into the North Atlantic on April 14, 1912. But Lord's book became a best seller and—more than that—established the floor for all *Titanic* books to come. It is where every *Titanic* scholar, writer, author begins, and many times ends. It is here that Lord's precious gift of being able to interview survivors of the *Titanic* is on display. He was able to capture the mother lode of personal recollections of that night. And that is the bulk of the book. The experience of the survivors is front and center, and here all the balloons are set flying that would populate every book to follow. Even the heavy-duty tomes on the *Titanic* have the breakers set up in the safe harbor of *Titanic* mythology. And so, like *The Wright Brothers*, where every book was based on the skewed Wright biography by Fred C. Kelly, little has deviated from Walter Lord's original thesis of how that night unfolded.

It's not that Lord's depiction of events is wrong. Far from it. It is valuable as a firsthand account of what the survivors went through, and in this way, it is a good resource. More, it is where Lord set his camera and the events he held up as important. You can hardly blame him. No one really knew what happened on the *Titanic* before it sank. Lord filled in gaps for a public hungry for the personal interest stories of the survivors. It was the ultimate voyeur's paradise. A you-are-there account of those 160 minutes from when the liner struck the iceberg to when she slipped under the icy millpond that was the Atlantic that night. The problem is that by focusing on these survivor stories told directly to Lord and repeated countless times in every book to follow, the other elements of that night are either ignored or treated to a quick glance before returning to the plight of the doomed and the lucky survivors. And because of this, an airtight mythology was erected that kept everyone within the buoys that mark *Titanic* lore for more than a hundred years. But there is another story, and it has a shocking epilogue, which is this: *Everyone could have been rescued if human will had not failed.* I will say it again: *Everyone could have been rescued if it were not for human failing.*

The story we have been told is that the *Titanic* was out there in the North Atlantic all alone. That it was amazing anyone was rescued at all. The truth is that she wasn't alone, thanks to the advent of wireless telegraphy, and, more incredibly, there were ships crisscrossing the Atlantic at the time, and two were in visual sight of the *Titanic* as she was sinking. Both ships had come upon the ice field that *Titanic* had stumbled into, and eventually both were ten miles away or less. The story of the *Titanic* is not one of catastrophe in an isolated oasis of darkness. Captain Smith and others saw the lights of ships on the horizon and directed passengers in the lifeboats to row toward them as well as Morsing the ships and sending up rockets to catch their attention. *Titanic*'s fate was not preordained, it was ordained by the failings of men in critical moments.

But let's start with the *Titanic* myth that has been handed us. It goes like this. The band played "Nearer My God to Thee" while the White Anglo-Saxon Protestant inheritors of male privilege, nay, the titans of their time, the Guggenheims, the Vanderbilts, gloriously threw down the gauntlet for all that was good and decent in the world of White Male Privilege and saw off their wives and daughters in lifeboats before having a final cigar, a dash of brandy, dressed in their finest like gentlemen, and when the icy Atlantic came for its due they shook hands all around and stepped off the last good ship of the Gilded Age and plunged to their icy fate to be forever memorialized in song, books, and film, and then stapled to the cultural moniker of all that was decent in the good old Edwardian world. Even Captain Smith does a last-minute cameo rescuing a baby and sending off his men with the admonishment, "Be British,"[2] before swimming up to a boat that is full and then telling the men, "All right boys, good luck and God bless you,"[3] then swimming off into the glorious mythology of the sinking of the *Titanic*. It does make for a great movie.

The mythology started right away. The *New York Times* headline on April 19, 1912, anchored *Titanic*'s place in the pantheon of the heroic. "744 Saw *Titanic* Sink with 1595. Her Band Playing; Hit Ice Berg at 21 Knots and Tore Her Bottom Out; "I'll Follow the Ship Last Words of Capt. Smith Many Women Stayed to Perish with Husbands."[4] The story of lone surviving *Titanic* Marconi operator Harold Bride also ran on this day under, "Thrilling Story by *Titanic*'s Surviving Wireless Man."[5] What was not to like here. The truth was Captain Smith never said anything close to *I'll follow the ship*; in fact, what is notable is Captain Smith not saying much of anything at all and then disappearing. A few women stayed with their husbands, but most got into the lifeboats and left. One could not blame them. But the Great White Christian ideal demanded the sinking of the *Titanic* be painted in the orange hues of a Swinburnian sunset, with women standing by their men and Captain Smith standing on the stern toward the end proclaiming, "I'll follow the ship!"[6] The only truth in the paper was that Marconi operator Bride's story *was* thrilling, and it was the first true account of the sinking.

The real story on board the *Titanic* is one of straight-up survival and a race to rescue people stuck on a giant ship that would sink in less than three hours. It is more of an Ayn Rand novel than one by E. M. Forster, depicting the primitive, naked, primal impulse to survive rather than the gilded patina of heroic Episcopalian men who suddenly grew a conscience after exploiting millions during the post–Civil War industrialization of America. Just ask the third-class passengers whom nobody bothered to interview after the sinking, mostly because a large number of them (537 out of 709)[7] never made it to the half-filled boats of first-class passengers, and, if they had, they would have been met with an oar or the sight of the stern receding from their view as the wealthy of America beat it away from the struggling masses thrashing in the water, fearing they might be swamped by this unwashed humanity. Is there a better metaphor than wealthy people sitting in their evening dress in boats half empty while poor and middle-class people who would have ended up in their factories drown all around them? If one took all fifteen hundred who perished and put their income against just a few of the first-class passengers, it would not even be close. First-class passenger John Jacob Astor IV was worth $150 million, or $3.7 billion today, while Benjamin Guggenheim came in at $95 million, and Isidor Straus, the man who created Macy's, was worth $50 million.[8]

The people on the *Titanic* saw their fate. The ship was being lifted into the sky with its bow gulping water and the propellors emerging as giant windmills from the ocean. Even before the ship split in two, people knew what was in store for them. The icy dark Atlantic was death. Period. So, people began to fight for survival. Stories have leaked out. Officer Lightoller brandishing a gun and firing to keep people away from the last lifeboat. Other officers shooting third-class men dead to keep them from the boats. People running along the ship and taking wild plunges into lifeboats that were descending. Crew gathering hands to keep men away from a collapsible lifeboat so that women and children might enter, then having to punch men who jumped in anyway. This is sandwiched in between the bread of self-sacrifice and Christian martyrdom.

So, the mythic Walter Lord patina of the heroic patrician class bravely facing death must still be scratched away to find out who the heroes of that night really were and who were the villains, or at least the unchivalrous, even as underwater robots plumb the depths of the Atlantic over *Titanic's* corpse. It is fitting that as of this writing there is a plan to bring the wireless cabin of the *Titanic* to the surface and find out if there are any secrets in the Marconi gear that wireless operators Jack Phillips and Harold Bride labored over until literally the water was sloshing around their ankles and the room was tilting toward a forty-five-degree angle. For if we are going to sweep away the detritus of mythology, then we have to look at the nuts and bolts of what people did do to try to save the 2,229 (exact number not known) people of the RMS *Titanic*. And more importantly, *what they didn't do.* And one could start with that decaying wireless shack three miles down on the floor of the Atlantic Ocean.

And if we are in the business of breaking up myths surrounding the night the *Titanic* sunk, then we might as well put this out there. It was a success. This goes against all the known thought, theory, extrapolation, judgment, revisionism, books, testimonials, movies, dreams. The success was that 705 (some sources have 701–713) people were pulled from the icy Atlantic because a little Italian named Guglielmo Marconi started putzing around in his parents' attic with the strange belief that electro-magnetic waves would be capable of carrying information from one place to another. Absurd thinking for the late nineteenth century, but when he sent some waves over a hill and the resulting shotgun blast came back that they had been received, that thinking changed forever, and the darkness of the North Atlantic or any sea, for that matter, would now be forced to give up her ghosts.

Now, it was a depressing failure of human dimension as well. Fifteen hundred and seventeen souls that were lost tell us this, and here is the smoking gun that punches a gaping hole in the building blocks of the Walter Lord edifice. Those 1,517 people who lost their lives in the darkness of April 15, 1912, constitute the millstone of human failing that hangs around the neck of Edwardian chivalry that has been the ruling trope of the *Titanic* mythology. There was a rescue operation, and a third of the passengers on that luxury hotel on the seas were rescued because of that operation. And what we don't know is that it was the first international rescue mission, truly a global effort to get to the mighty leviathan before she sank beneath the freezing water surrounded by icebergs.

One hundred and sixty minutes. That is all the time rescuers would have before the largest ship in the world slipped beneath the icy Atlantic. Truly, this bridge of two hours and forty minutes was the dividing line between the Edwardian era and the industrial mechanization of the twentieth century. When Captain Smith told his two young Marconi operators as the ship was slipping into the sea, "Men you have done your full duty. You can do no more. Abandon your cabin. Now it's every man for himself,"[9] this announcement would unwittingly mark the collapse of the old world and define the ethos of the new. It would sadly define the sinking and the rescue and would be the dark palette for the few stars who would become the treasured footnotes of human exceptionalism.

There was amazing heroism and astounding incompetence and cowardness against the backdrop of the most advanced ship in history sinking by inches with luminaries from all over the world. It is a story of a network of wireless operators on land and sea who desperately sent messages back and forth across the dark, frozen North Atlantic to mount a rescue mission. More than twelve ships and four countries would be involved in the rescue of *Titanic* survivors. At the heart of the rescue are two young Marconi operators, Jack Phillips, twenty-five, and Harold Bride, twenty-two, tapping furiously and sending electromagnetic waves into the black night as the room they sat in pitched toward the icy depths and not stopping until the bone-numbing water was around their ankles. Then they plunged into the water, having coordinated the largest rescue operation the maritime world had ever seen, thereby saving 705 people by their efforts.

We don't understand what a leap forward Marconi's invention of wireless telegraphy was for people in the late nineteenth and early twentieth centuries. Up until the young twenty-two-year-old Marconi astounded the world by sending messages across the English Channel, people on ships were isolated, and people on land had to depend on a telegraph wire that broke many times in the far reaches of the expanding country. But Marconi had demonstrated that electromagnetic waves could travel vast distances, and this changed maritime transportation for good. Ships just vanished before the wireless age. No trace was left and many times no survivors. But now a ship could let another ship know if she was in distress or pass on the information to a land-based station that could relay it up the line. When the *Titanic* was sinking, her sister ship *Olympic* relayed the information, and a Wanamaker department store employee in New York, David Sarnoff, on the roof of the store in a Marconi wireless room plucked the signals from the air and let the world know.

The race to save the largest ship in the world would begin at 11:40 p.m. on April 14, when the iceberg was struck, and would end at 2:20 a.m. on April 15, when her lights blinked out and left 1,517 people thrashing in 28-degree water. Although the race to save *Titanic* survivors would stretch on beyond this, most people in the water would die, but the amazing thing is that of the 2,229 people, 705 did not, and this was the success of the *Titanic* rescue effort. The only reason every man woman and child did not succumb to the cold depths is because of Phillips and Bride in the insulated telegraph room known as the Silent Room. As the ship took on water at the rate of four hundred tons per minute from a three-hundred-foot gash, these two men tapped out CQD and SOS distress codes that would inaugurate the most extensive rescue operation in maritime history, using the cutting-edge technology of the time—wireless.

The 160 minutes of *Titanic*'s sinking would forever change the technology into a lifeline for those who would be rescued from the frigid lifeboats and a tragic footnote of its infancy for the 1,517 who would perish. Guglielmo Marconi, who would meet the surviving operator, Harold Bride, at the pier in New York as the *Carpathia* docked, would be hailed as a hero for his lifesaving technology and then scorned when it was found that Bride and others were told to hold their information so Marconi might strike a deal with the highest bidder, which turned out to be the *New York Times*. This new technology, used in the biggest disaster of modern times, created thorny ethical questions.

The race to save the *Titanic* was a knight's mission in the dead of a freezing night on the Atlantic. It was two men huddled in the transmitting room sending out their plea for assistance with distant ships snatching the meaning and interpreting those dots and dashes in a myriad of different ways, relaying it on to a shocked world. It was the actions in Britain and New York and the ships steaming toward the stricken ship of dreams with the only thought to reach her before people succumbed to the icy water. The natural drama of *Titanic*'s death throes can be juxtaposed against this race against time by the other ships and the tragic consequences of missed opportunities.

The sinking is a Greek tragedy of epic proportions that would play out for Captain Stanley Lord of the SS *Californian* and Captain James Henry Moore of the *Mount Temple*, and this same drama would reward Captain Arthur Henry Rostron of the RMS *Carpathia*. The *Titanic*'s sister ship, *Olympic*, would do an about-face and immediately light up every boiler in a desperate attempt to reach the sinking behemoth. The *Baltic*, *Virginian*, *Franz Albert*, *Mauretania*, *Parisian*, and *Cincinnati* would all make valiant efforts to reach the *Titanic* from varying distances while assisting in the relaying of messages from Marconi operators Bride and Phillips. It is the new technology of wireless telegraphy that would intersect with Old World sensibilities to give us our first modern tragedy, showing the miracle of wireless communication and its tragic limits. The SOS or CQDs were the veins of the extended heart that encompassed the rescue ships and their relaying information to farther stations on shore. In this way the *Titanic* was the first real-time tragedy played out with all the misinformation and heartbreak that is the result of an early developing form of communication.

In the end, nineteenth-century gallantry would be swallowed by a very twentieth-century fight for survival. The story of the rescue effort has the grace and pathos of the mythical musicians who played until the ship slipped under water and the unsinkable Molly Brown, who demanded the half-filled lifeboats return to take on the people in the water. It is a buried story of men laboring through the night knowing time was not on their side and that they were slipping toward oblivion with every tap of their Marconi key. It is the hidden stories of the ships, their captains, crews, wireless operators, owners, newspapers, Marconi himself, all laboring under impossible circumstances for a doomed vessel, and yet one more branch of the *Titanic* story that continues to fascinate us. The rescue of 705 people is a success story, but it is the might-have-been of the 1,517 souls that keeps us coming back for more. So, let's go back to that night, April 14, 1912, that author Walter Lord deemed *A Night to Remember*. Indeed, it was.

SHIPS INVOLVED IN THE RESCUE OF THE *TITANIC*

Californian	10 to 18 miles from *Titanic*
SS *Parisian*	45 miles southwest of *Titanic*
Mount Temple	50 miles southwest of *Titanic*
RMS *Carpathia*	50 miles southeast of *Titanic*
SS *Birma*	70 miles southwest of *Titanic*
SS *Frankfurt*	140 miles southwest of *Titanic*
SS *Virginian*	170 miles west of *Titanic*
RMS *Baltic*	200 miles east of *Titanic*
RMS *Olympic*	400 miles southwest of *Titanic*
RMS *Caronia*	700 miles from *Titanic*
SS *Prinz Friedrich Wilhelm*	120 miles south of *Titanic*
SS *Cincinnati*	550 miles from *Titanic*

PROLOGUE

December 12, 1896, London's Toynbee Hall

GUGLIELMO MARCONI WAS NERVOUS. HE WAS AN INVENTOR PER SE, BUT more than that he was a perfector of science for practical applications. He was only twenty-two and looked like a boy. He took other ideas and concepts and then saw how they could be useful for others as well as himself. He had been working tirelessly to perfect the transmission of Hertzian waves that Oliver Lodge had demonstrated in an 1894 lecture on Hertz at the Royal Institution. Lodge claimed Hertzian waves could travel only a half mile. Marconi claimed he had sent the electromagnetic waves a mile. Then he sent the waves from one rooftop to another, almost two miles. Newspapers wrote articles about the young Italian who had invented wireless telegraphy and called the Hertzian waves "Marconi waves." It was time for a public demonstration.

William Preece, the chief electrician for the British Post Office, would give the lecture where Marconi's wireless telegraphy would be demonstrated in a simple yet startling manner. Preece was preeminent in the new field of British telegraphy and was well known for his lectures. This would be one of Preece's barn burners, displaying the invention of the elfin Italian who had shown up in his office with his amazing wireless invention. The hall was packed as Marconi and Preece set up the demonstration. Marconi was fearful someone might steal his idea before he could patent it, so he kept his receiver and transmitter hidden from view in painted black boxes. Marconi put the transmitter box on the podium and then walked across the room and put the receiver box on a chair. Preece began the lecture.

Preece was an august man with thick glasses and a large bushy white beard, which was the fashion for learned men. Preece explained his own efforts to send signals across bodies of water, which had not amounted to much. He then turned to Marconi and announced he would demonstrate an amazing new breakthrough by a young Italian inventor. This sort of buildup was in line with magic shows of the time. Preece walked up to the box as the murmuring in the hall fell silent. A shot rang out and the audience jumped. It was the sound of the spark jumping the contacts as Preece pressed the key. The audience jumped again when the bell inside the receiver box on the far side of the room rang. Many in the audience had attended late nineteenth-century magic shows, and this seemed to be more of the same. A magician, Nevil Maskelyne, had been making the rounds and performing at the Egyptian Hall in Piccadilly. So, in the world of magic this was rather tame—much more dramatic to see a woman sawed in half than a box that rings.

The people in audience looked for the wires. Electricity was a known quantity, and wired telegraphy was in great use. There had to be wires. Preece split the air again with a loud crack and the bell dinged again. People looked at the floor. There had to be wires somewhere. But now Marconi walked up to the black receiver box and picked it up from the chair with a flourish. He then began walking around the hall while Preece lit the darkness with a blue spark and the bell dinged in front of Marconi, who carried the box like a priest presenting a holy relic. *Ding, ding, ding, ding.* The audience stared openmouthed. There were no wires behind Marconi, who walked up and down the aisles in a black tuxedo like a magician with his newest trick, wireless telegraphy—as the bell dinged and dinged and dinged and dinged. But it was no trick. It was the call of a new century, a new world.

Marconi lost no time setting up the Marconi Wireless Telegraph Company with backing from his Anglo-Irish cousin Henry Jameson Davis. The company focused initially on experimentation to perfect Marconi's newly patented equipment, but Marconi was as much a salesman as an inventor, with a knack for seeing the most lucrative application of this new technology. He focused on maritime application of wireless, and by the turn of the century he had patented a tuning application that allowed senders to communicate on different wavelengths. Now stations or ships could send signals to each other and, in theory, not have any interference. Marconi's patent No. 7777 in 1900 set the stage for his dominance in this new technology, which was akin to the advent of the internet.

He quickly formed the Marconi International Marine Communication Company and outfitted the German liner SS *Kaiser Wilhelm der Grosse* with Marconi equipment in 1900. The race was on to put Marconi equipment on as many ships as possible and to staff the ships with Marconi operators who answered only to the Marconi Company. It was a young man's game, with the company looking for men between the ages twenty-one to twenty-five with experience in inland or cable telegraphy. The ability to be able to receive and send Morse code at a rate of twenty-five words a minute with a Morse key was required. Applicants were then sent to Marconi's Liverpool training school for courses in wireless telegraphy that would allow them to take the examination given by the postmaster general. If they passed, they received certificates of competency. Operators were expected to be able to repair their own equipment and to understand the new technology they were operating. To that end, there were courses in "elementary electricity and magnetism; fundamental principles of wireless telegraphy . . . how to trace and remove faults and repair breakdowns . . . clerical work in connection with telegraphic accounts and returns and general routine and discipline on board ship."[1]

The operators held the rank of an officer on the ships. The rate of pay was slightly higher than land-based telegraphers and slightly higher than railway telegraphers. But the life of the wireless operator on the seas was exciting, as young men had opportunities to travel the world and worked literally on the cutting edge of the information age. At night, the Marconi operator became the conduit to the world. No one quite understood why, but signals that could only go three hundred miles during the day leapfrogged two thousand miles at night. This was due to an ionized layer in the atmosphere that bounced the electromagnetic waves back to Earth. The sunlight distorted this effect, but at night true long-distance wireless took place, and this brought the advent of shipboard newspapers.

News passed on by shore stations to ships was then transcribed into a publication that passengers could read in the morning over their baguettes and coffee. It was amazing to read about the weather, sports, and world events in the middle of the ocean. The young operators took pride in being the ones who deciphered the wireless Morse code that ushered in this modern world. Few on board understood the technology or the new role wireless would now play in the maritime world.

When the largest ship in the world left for her maiden voyage, it left equipped with the most powerful Marconi equipment, which would allow her to stay in contact the entire way. *Titanic* had four parallel aerial wires two hundred feet high and six hundred feet apart transecting her funnels to pluck signals from the air and a 5 kW motor generator set yielding three hundred volts to send her voice to the farthest shore stations. Her daytime range was four hundred miles, and the nighttime range was two thousand miles.

The *Titanic* had backup power to spare with the motor of the system powered by the ship's 110-volt DC circuits but with an independent oil-driven set on the top deck and a battery of accumulators provided as standby. This was a state-of-the art ship with state-of-the-art communication equipment to stay in touch on the North Atlantic. No one dreamed this technology would shortly be needed in a race for survival.

CHAPTER ONE

New York, April 18, 1912

OUT OF THE FOG CAME THE SHIP LIKE A GHOST. HER BOW LIGHTS emerged first, and then the glittering lights high up on the bridge. A foghorn announced her arrival as the steamship glided up the Ambrose channel toward the waiting world. Brilliant magnesium flashes from photographers on boats framed the RMS *Carpathia*. The blackened soot from her funnels roiled into the moisture-laden air. Launches hired by newspapers to get the first quotes from survivors of the *Titanic* bobbed around in the swell generated by the lantern-lit ship. The smaller boats maneuvering in the whitecapped wake of *Carpathia* jockeyed for position as reporters shouted questions promising hundreds of dollars for a quote. A voice rose over the thrum of the steam engines and the echolalia of men shouting with megaphones.

"Are you one of the *Titanic* survivors?"

A woman shouted back.

"Yes," the voice replied hesitantly.

"Do you need help?"[1]

The reporters were hanging on like monkeys to the railings of the launches bobbing in the waves created by *Carpathia*. From somewhere along the railing of pale faces on *Carpathia* came the voice again.

"No!"

"If there is anything you want done it will be attended to," the reporter shouted in the closest launch, daring fate by coming perilously close to the giant liner.

"Thank you," came the reply.[2]

The newsmen became more desperate. The boats were perilously close to being crushed by the *Carpathia*. They waved fifty-dollar bills in the air. "Jump overboard," they cried to the *Titanic's* surviving seamen. "We'll pick you up!"[3] The world had been waiting for word from the liner that had gone strangely silent. President William Howard Taft could get no response from the wireless operators aboard ship concerning his aide Archie Butt. The president had then sent the cruiser *Salem* to meet the *Carpathia*. This only increased the frenzy in the rough chop and cold drizzle as reporters climbed aboard the pilot ship pulling up to *Carpathia* to take her into the harbor. Even the mayor and a group of officials in their launch were unsuccessful in gaining access. Whistles and steamship blasts sounded across the harbor. Reporters tried to jump on board the ship as the harbor pilot entered *Carpathia* to take control of the ship. Third officer Rees punched a reporter in the mouth, knocking him back into the pilot boat.

"Pilot only,"[4] he bellowed.

More magnesium flashes exploded as the pilot took control, and the great misty liner headed toward the Cunard pier then veered off and stopped at the White Star dock. She unloaded the thirteen empty *Titanic* lifeboats with the solemnity of an undertaker delivering bodies. It was all that was left of the greatest ship ever built. The liner turned then and headed back to the Cunard pier. Soggy and cold, the crowd of thirty thousand people stood grimly in the New York night along the east bank of the Hudson River. Lightning flashed again, setting the stage for the return of the only survivors of the *Titanic*. It was the moment the world had been waiting for in rapt attention for three days: the return of the *Carpathia* with the survivors of the *Titanic* on board. New York was a cold, drenched, glittering city from the deck of the ship that had risked all to rescue the 705 people freezing in their lifeboats in the Atlantic. Captain Rostron had watched from the bridge as he approached New York and was amazed at the site of the boats and people waiting for the arrival of his ship.

As we were going up the Ambrose Channel, the weather changed completely, and a more dramatic ending to tragic occurrence it would be hard to conceive. It began to blow hard; rain came down in torrents and to complete the finale, we had continuous vivid lightning and heavy rolling thunder.[5]

The lightning did not faze the sharply dressed man standing in the rainy darkness impervious to the cold and rain. He had seen lightning many times before. He made lightning when he sent the first wireless signals across the Atlantic Ocean. Nobody said he could do it. They said the Earth was curved and the electromagnetic waves would simply go off into space. He had generated so much current to send the signals that the keys emitted great arcs of blue light and cracked like thunder. *S* was the first letter to traverse the Atlantic from England to Newfoundland.

Guglielmo Marconi stood along the pier as the passengers disembarked from the RMS *Carpathia*. He stood in his tight Italian suit with his dark hair painted down with oil and his high white collar pinching his neck. The rain and cold were irritating, and the squinting was giving him a headache. He watched as survivors from the *Titanic* came down the gangplank—some on stretchers, others assisted by nurses, some walking off as if they had just come from a holiday. The crush of reporters, nurses, the Red Cross, the curious, political figures, police, doctors, lawyers, relatives was overwhelming. Marconi ignored it all, even as reporters recognized him and were pushed away by his manservant.

He was there to intercept the one man whom every reporter would give his left arm to interview, Harold Bride. He was the only surviving Marconi operator on the RMS *Titanic*, who, along with Jack Phillips, had stayed at his post orchestrating the rescue of the doomed ship even as the water entered the telegraph room. Not only was he a genuine hero, but he also knew things about the sinking of the largest ship in the world that no one else knew. He himself was now the story of the unsinkable luxury liner *Titanic* hitting an iceberg and sinking on her maiden voyage with catastrophic loss of life. This made him the most valuable man on *Carpathia*, and Marconi had instructed him with many messages to keep his mouth shut and that he would handle his story. The message was cryptic and to the point. "Arranged for your exclusive story for dollars in four figures. Mr. Marconi agreeing. Say nothing until you see me."[6] Marconi understood the value of Bride to his company and how other competitors would love to get hold of him. It was imperative that Marconi get to him first and take control of the young man who had witnessed the most tragic event of the very young century.

The amazing thing was the thirty-seven-year-old inventor of wireless telegraphy and founder of the Wireless Telegraph and Signal Company was supposed to have been on the *Titanic* with his wife, Beatrice. He had canceled and sailed on the *Lusitania* (another doomed ship) because he needed a first-rate stenographer for the voyage. His wife did not cancel until the last minute when their son Giulio had contracted a high fever. It was a close shave, but Marconi chalked up his luck to the same divine intervention that made him a wealthy young man at age twenty-five when he went against the conventional wisdom that said electromagnetic waves could never traverse the ocean because the world was round, and they would simply fly off into space. Marconi, who was not a scientist of any kind but rather a gifted entrepreneur who had stumbled into wireless telegraphy through trial and error, saw no reason people shouldn't be able communicate across the oceans. His ignorance had worked in his favor, as it made him believe Hertzian waves could do things that everyone said was impossible. He was part entrepreneur, part gambler, part vision-ary, and part dilettante. But he saw the enormous potential very early to become rich off the new technology, and before long he was linking ships to land-based stations and each other with his operators and his equip-ment. He had a nose for profit, and that nose had put him on the dock to greet his one surviving employee of the RMS *Titanic*.

Marconi put his hand over his brow and motioned the men with him to go up the gangplank. A young reporter, Isaac Russell of the *New York Times*, followed the famous inventor along with another representative of the Marconi Company as they walked past the silent disembarking survi-vors. Marconi went directly toward the wireless room on the ship, notic-ing the antenna strung up on the mast. Several officers approached him and then begged off when he identified himself. As Russell later wrote,

What mortal power could issue orders to bid Marconi stop? Sailors fell before us. Eyes popped out and lips froze with one word, "Mar-

coni," half uttered upon them. Gaping guards to the right of us, gaping guards to the left, and gaping guards in front of—and beside themselves and all ready to die—to see that Marconi passed in spite of every order they had received. . . . The magic word had travelled along—"Marconi" came up in a murmuring mutter from the guards ahead.[7]

Marconi weaved through the labyrinthine ship and entered the dimly lit wireless room that smelled of electricity, a slight burning scent one could only compare to as the smell of dust in an old fan. There they "found a boy sitting on a high stool—sending, sending, sending. . . . On his wireless stand before him sat a plate of dinner all uneaten. On the wall of the cabin hung a photograph of Marconi."[8] The slight young man sat with his right arm extended pressing down on the key with the blue spark dancing from under his fingers. Dark crescents painted his eyes, and his ghostly pallor made Marconi think of his dead father. The staccato cracks of the electric spark jumping the contacts filled the room. Bride wore a large overcoat that swallowed him up to his severely frostbitten feet, which resembled bowling balls wrapped in white bandages. He looked like death, and rightly so; he had seen more death in a few hours than most people would see in a lifetime. His skin was tight, his damaged feet were propped up on a chair. He did not see Marconi or the men with him. To reporter Russell he looked like a child who had just been through a war and having survived had become almost saintly. Russell felt a tickling down in his gut. It was Harold Bride, the only surviving wireless man of the RMS *Titanic*.

Marconi, who had set up the school to train Bride and had outfitted the ships of the world with wireless, listened for a few moments and deciphered the message in his head that Bride was sending. He understood that Bride had been assisting the *Carpathia* wireless operator in sending lists of survivors. He also understood that Bride had received his admonishment to "keep your mouth shut and hold your story, it is fixed for you so you will get big money,"[9] as he had secured a deal for a couple thousand dollars if he gave his story first to the *New York Times*. Marconi cleared his throat and pinned down the message Bride was working on with his finger.

"That's hardly worth sending now, boy."[10]

Bride looked up. His eyes flared recognizing the famous inventor who quickly grasped his hand and held it warmly. Bride protested. "The people out there they want these messages to go. I must send them, the people waiting by the cabin."[11] Marconi explained everyone had gone ashore. "It took what seemed a long time before the recognition came into the young man's eyes, but when it did, he even smiled a little. 'You are Mr. Marconi,' Bride finally said as he took his fingers from the telegraph key."[12]

"Mr. Marconi," Bride said. "Phillips is dead. He's gone."[13]

The reporter from the *New York Times*, Russell, whipped out his notebook and pencil. Marconi stared at the pale young man. It was the story of the century. The sinking of the *Titanic* was an incredible story, but the rescue attempt was a thrilling story as well. Marconi's employees had been at the nerve center of this amazing rescue drama and inaugurated the saving of more than seven hundred people from a watery grave by their gallant actions. They alone knew what the captain did, the crew, how fast the ship sank, the heroes, the cowards, what ships responded and what ships did not. The *Carpathia* was there because of this wan young man. Every *Titanic* survivor getting off the ship owed their life to Harold Bride, yet most would never hear his name.

Bride put on his hat and buttoned up his overcoat. The two men with Marconi carried him through the ship and to the gangplank, where a photographer snapped a picture. In the picture, Bride is literally hovering in the air with his two clubbed feet out in front of him, a fedora pulled low on his head. The men then carried him to the car, where Marconi slid in next to him, pulling the curtains.

Reporters tapped on the windows and shouted through the glass. Marconi shook his head and brushed lint from his pressed black trousers and turned to the young man, who looked like a frightened cherub at that moment. Bride was in shock. He had been busy on the *Carpathia* assisting wireless operator Harold Thomas Cottam in communicating with relatives and the White Star officials as well as officials in New York. But Marconi had given him strict orders to tell no one of what he had seen. He had not even responded to the U.S. Navy or the president who wanted to know if his personal assistant Archibald Butt had perished. That was more a matter of him being busy with passenger messages and the lists of who had survived and who had perished. Also, the navy telegraphers were awful and didn't understand International Morse. So, he had said nothing to anyone and helped Cottam with the workload and nursed his frostbitten feet.

They entered the Strand Hotel, where the editor of the *New York Times*, Carr Van Anda, had "hired an entire floor of the Strand Hotel, located at Fourteenth and Eleventh Avenue, just a block away from where the *Carpathia* would dock and outfitted it with four telephone lines with direct connections to the *Times* rewrite desks."[14] Van Anda had already scooped all the other papers by declaring the *Titanic* had sunk before anyone was really sure. He was not going to miss out on the biggest story of the young century now. Bride was whisked through the lobby surrounded by several large men and Marconi. Bride was then ushered down the hall to a suite where he entered and found a room full of men and cigar smoke waiting for him. They stood as one as he entered, and Marconi stepped forward as Bride sat down slowly on a plush couch.

Bride felt his face warm. Jack Phillips had sent most of the distress calls, but he had perished in the Atlantic. The men stared at the small youth in front of them. Such a colossal disaster of unimaginable proportion. This gigantic ship that was said to be unsinkable and yet she had sunk and taken more than fifteen hundred souls with her and left this urchin of a man, all of twenty-two, to tell her tale. It was simply not believable, but it was the story of the century and it was here in this room. What had this man seen and what did he know about this unbelievable disaster? The *Titanic* had sunk and was gone, and this was the man who had tapped out her sinking to the very last, sending out sparks into the cold darkness that had brought ships speeding toward her from all over the sea. Yet only one ship had reached her in time to save anyone. Reporter Isaac Russell sat down in front of him.

The men smoked and stared at Harold Bride as if they just had a large dinner and all were relaxing over cigars. His eyes went down to his heavily bandaged feet that had frozen in the icy water of the Atlantic. Bride cleared his throat. The whole sinking of the *Titanic* had the eerie feel of a dream, a nightmare that did not seem real. As many times as it played out in his mind, it did not seem believable that this colossus of modern technology had sunk and carried off all those people with her. Bride had kept it from his mind and stayed busy on the *Carpathia*, not thinking about the death of Phillips who had stayed in the Silent Room with him until the very end. He had not thought about any of it, but now it was crashing in on him like the water that swelled toward him when he stepped out of the wireless room and carried away Phillips and then himself; that icy water that felt like a thousand knives and took his breath away and made his heart feel as if it would tear out of his chest.

Bride stared at the men smoking cigars with legs crossed, shoes shined. Their beards were immaculate, their demeanor one of order and continuity. How to convey what had happened just four days ago out on the dark frozen seas of the Atlantic. He stared down at his feet that throbbed and felt like tacks were being driven into his toes. He then looked at the men again. He had survived by luck under an overturned boat and then had been pulled on top of it. Sheer luck. That was the only reason he was here. Sheer luck.

Bride swallowed, nodded once. He had just taken over and put on the headphones. Jack Phillips was tired and heading for his bunk in the adjoining room. Bride would always remember Captain Smith's eyes when he entered the room and looked at him. That bearded giant of the seas had just come to the realization that his ship was sinking. Bride looked up at reporter Russell, who would later reveal that when he typed up the story, he cried. "I saw that the ribbon was going wrong and spreading ink about and became aware that tears were falling on the paper."[15]

Harold Bride then began to speak in a low voice.

"I was standing by Phillips, telling him to go to bed, when the captain put his head in the cabin." He looked up at the men. "We struck an iceberg . . ."[16]

CHAPTER TWO

Magic and Dreams

THERE WAS A MOMENT OF MAGIC AND DREAMS. THE EDWARDIAN world teetering already on the knife edge of the twentieth century was summed up by a second-class passenger, schoolmaster Lawrence Beesley, in a book written eight weeks after the sinking. It was a world still of libraries and letter writing. On the afternoon of Sunday, April 12, Beesley found it "too cold to sit out on deck to read or write, so that many of us spent a good part of the time in the library, reading and writing. I wrote a large number of letters each day and posted them day by day outside the library door."[1] The shipboard life had a placid, even staid, rhythm. "Each morning the sun rose behind us in a sky of circular clouds, stretching round the horizon in long narrow streaks and rising tier upon tier above the skyline."[2] Titanic passengers drowsed their days away in the ornate Victorian rooms. Beesley's description of life on that Sunday afternoon has the tenor of a country house in England.

> I can look back and see every detail of the library that afternoon—the beautifully furnished room, with lounges and armchairs, and small writing or card tables scattered about, writing bureaus round the walls of the room, and the library in glass cased shelves flanking one side—the whole finished in mahogany relieved with white fluted wooden columns that supported the deck above. Through the windows there is covered corridor, reserved by general consent as the children's playground and here are playing the two Navratil children with their father.[3]

Beesley gives us this yawning afternoon before the storm and sees a "man and his wife with two children, and one of them he is generally carrying; they are all young and happy; he is dressed always in a grey knickerbocker suit—with a camera slung over his shoulder."[4] Then an ominous comment. "I have not seen any of them since that afternoon."[5] He describes the library steward as "thin, stooping, sad faced, and generally with nothing to do but serve our book; but this afternoon he is busier than I have ever seen him, serving out baggage declaration forms for passengers to fill in."[6] Beside him are "two American ladies, both dressed in white, young, probably friends only."[7] They are talking to a photographer from Cambridge and his young French wife. In the center of the library are two Catholic priests, "one quietly reading—either English or Irish and probably the latter—the other dark bearded with broad brimmed hat, talking earnestly to a friend in German. . . . I can remember only two or three persons who found their way to *Carpathia*."[8]

The Wright brothers had flown nine years before, and science was in the forefront with vaccines, X-rays, and then something that had enhanced voyages on luxury liners—wireless telegraphy. Guglielmo Marconi had changed the nature of the isolated voyage across the ocean by adding a wireless room also known as the Silent Room (named for the insulation that muffled the loud cracking of the arcs sending out Morse signals and walling out sounds so the operator could hear the incoming signals) to every ship of the line, and now passengers could let someone in New York know they were approaching and having a fantastic time on the largest, most luxurious ship in the world, the RMS *Titanic*.

That a crack of the spark transmitter unleashing its charge from the insulated room that jumped the ionosphere between two contacts and thereby sent either a dot or a dash to be decoded in the Cape Race Station in Newfoundland and then relayed on to New York or London was simply amazing. The *Titanic* with its large four-wire antenna slung between the two masts like giant telephone pole wires had the most powerful set ever put on a ship, 1.5 kW, boosting a range of two thousand miles at night.

Gone was the isolation on the high seas where ships slipped out of contact for weeks and many times for good. The inboxes of the wireless operators quickly filled with passenger requests even as warnings of ice fields came in. *Titanic* had received six warnings from ships passing through ice fields on April 11th, five on the 12th and three on the 13th. April 14th had seen seven more warnings from ships in heavy ice. These messages were logged, which gave the longitude and latitude of each sighting, and passed on to the bridge where the captain and other officers took appropriate actions. If "put together, the six messages indicated an enormous belt of ice stretching some seventy-eight miles directly across the big ship's path."[9]

But no one put the warnings together. In fact, there was no protocol for ice warnings at all. The Marconi operators made a decision which messages went to the bridge and which could languish in the receiving box in the wireless room. Even when the messages reached the bridge there was no clear procedure. "If the recollections of the four surviving officers are any guide, most of the warnings went unnoticed on the bridge."[10] The two operators kept the radio in operation twenty-four hours a day by working in shifts. Wireless was unregulated and protocol was lax. There were no official rules governing what many still viewed as a novelty in 1912. Sometimes a note was affixed to the chart showing the ship's progress, sometimes a warning found its way into the captain's pocket, which happened with a warning from the *Baltic*. It was there when Captain Edward John Smith went to lunch at 1:30 p.m., where he handed it off to the managing director of the White Star Line, J. Bruce Ismay, who pocketed it and then showed it around to Mrs. Thayer and Mrs. Ryerson, two first-class passengers the director wanted to impress. Ismay held onto it until the evening when he bumped into Captain Smith coming out of the smoking room and gave it back to him. Clearly the path for an ice warning was circuitous at best.

But in the wireless room there was protocol. Jack Phillips sat hunched over in the wood-paneled wireless room with his right hand on the telegraph key with the headphones pinching his ears slightly. He had developed a crick in his neck and a strange fluttering sensation in his wrist from tapping the key. He listened to the long buzz, then short buzz, then a long buzz followed by two short buzzes. The modern ear would hear a slightly fuzzy sound like static on an old radio. The thrum of the 52,310-ton ship plunging through the North Atlantic was lost on Phillips as he concentrated on the next incoming message.

The reciprocating steam turbine engines capable of producing 50,000 horsepower resonated through every part of the ship. This was the comfortable backdrop, the vibration in a cup of tea, the slight shake of a waterglass, the throb rolling up through a pillow that let everyone know all was well on the ship. It was the heart of the ship that not only propelled the nearly 900-foot-long ship through the water at an incredible 22 knots but also supplied power for

> *evaporation and refrigeration plants, four passenger elevators, a fifty-phone switchboard, a five-kilowatt Marconi station, several hundred individual heating units: a gymnasium complete with all the latest electrical devices, eight electric cargo cranes of a combined lifting weight of eighteen tons, numerous pumps, motors and winches.*[11]

Phillips had finished deciphering the incoming message and sat back. It was another ice warning. They had started coming in on April 11th and he was tired. He had been sending messages for passengers all day and Harold Bride, the second operator, was soon to relieve him. Wireless was a young man's game, much like the early days of Facebook and Twitter, where few understood the technology. Phillips and Bride were low-paid employees of the Marconi Company and in fact reported to their superiors of the company rather than the captain. They were regarded by the ship's crew as a breed apart, working in a cutting-edge industry and using their own slang of "Old Man," or OM, in messages, a sort of English prep-school jargon. They knew the future belonged to them, and the first-class passengers had only recently discovered they could send missives from the middle of the ocean.

Phillips typed out, "Hello boy, dining with you tonight in spirit, heart with you always. Best Love Girl."[12] He shot this to the Cape Race Station in Newfoundland where it would be relayed to New York. Wireless was a patchwork quilt of stations on ships and land that hurled each message through the ionosphere as far as possible until the next station captured it and hurled it again. "Fine voyage, fine ship."[13] From Phillips's telegraphic key to Cape Race to New York's Fifth Avenue, it was nothing short of amazing in 1912. "Arrive Wednesday. *Titanic* maiden voyage. Meet me, vessel worth seeing."[14]

Indeed, Captain Smith, who was taking the *Titanic* across on her maiden voyage, regarded the ship as worth seeing. While not a skeptic of the new technology, he viewed the wireless messages as an augmentation to his thirty years' experience commanding ships on the high seas. The ice warnings that Phillips and Bride had been sending up to the bridge were duly noted (some), but the twenty-fourth boiler of twenty-nine had been lit and the *Titanic* was not slowing down as she edged closer to the reported ice along the parallel 41.50 north latitude. The paying trade in the young business of wireless at sea was customer messages; the ice warnings were passed dutifully along with the congratulations to Captain Smith for his securing the *Titanic* command and on his impending retirement. This was it for the old grand master of the sea. After the *Titanic*'s maiden voyage the bearded patriarch would retire, and this made the voyage more special for the crew and the passengers. Many of the passengers had followed Captain Smith from ship to ship, and he was known as a congenial, charismatic captain who could play both the role of ambassador of goodwill for the Cunard Line and the captain of the largest ship afloat. No small thing.

But Phillips was tired. The messages had piled up when the wireless broke down at 11 p.m. on Friday night. He and Bride had stayed up half the night running down a short in one of the secondary circuits and finally got the wireless functioning again at about 3 a.m. So, he had been playing catch-up as the messages had poured in from the first-class cabins, where a Regal Suite booked for $4,350 ($260,000 today).[15] Even with the lack of a clear path for ice messages, the *Titanic*'s system for sending a wireless message for passengers was advanced for the time.

Passengers sent their messages at the enquiry office at the starboard side of the forward first-class entrance. The handwritten messages were paid for at the desk at the rate of 12s 6d for the first ten words and 9d for each additional work. Messages were then sent by pneu-

matic tube to the wireless cabin where the operator noted the word count and transmitted the message.[16]

When an incoming message came in, the process reversed itself. The message was transcribed, typed on a Marconi form, then sent back by pneumatic tube to the pursuer's desk, where a bellboy took it to the passenger's cabin. For 1912, this was a very modern system of communication. If the message pertained to navigation, then it went to the bridge, but curiously there was no formal procedure. Since Phillips was the senior operator, usually Bride took it up to the bridge, but sometimes the messages lingered on the wireless desk as the onslaught of passenger missives increased. The fact that the streamlined efficient method of transmitting messages belonged to people trying to get together for lunch in New York versus the haphazard walking of a message to the bridge to report on dangerous ice in the path of the ship shows that no one had yet understood the true mission of wireless telegraphy.

But the ice messages had increased during the last few days, with many on April 14th: *"West Bound Steamers report bergs growlers and field ice on 42N from 49 to 51 West"*[17] *"Amerika passed two large icebergs in 41.27 N, 50.8 W on the 14th of April"*[18] "SS *Mesaba* reported *"Lat 42 N to 41.25 Long 49 W to Long 50.30 W saw much heavy pack ice and great number large icebergs also field ice."*[19] A message from the SS *Baltic* was given to Captain Smith at 1:42 p.m. *"Greek Steamer Athena reports passing icebergs and large quantities of field ice today in latitude 41.51."*[20]

Phillips set the latest ice warning aside for Bride to take to the bridge. The wall heater pulsed against his leg as he began working the land station Cape Race. It was 5:30 p.m. on April 14th and the temperature had dropped rapidly outside. Passengers had vacated the decks as the strange cold put whiskers around the mounted lights. The wireless room was cozy with the bare candescent light overhead throwing a patina on the brass telegraph key and dark tuning knobs. A nearby ship, the *Californian* chimed in and nearly blew Phillips's ears off. He resisted snapping back as he took down another ice warning. "Six thirty PM latitude 42.3 N longitude 49.9 W Three large bergs, 5 miles to the Southwards of us."[21]

Phillips gave the message to Bride, who took it to the bridge, but Captain Smith would never see this last ice warning. He was at a dinner party given by Mr. and Mrs. George D. Widener with the personal advisor to the president of the United States, Major Butt. Phillips continued with his work hunching closer to the heater, thinking of his bed waiting for him after Bride relieved him. The steady vibration of the engine deep in the bowels of the ship was turning the three huge propellors at seventy-five revolutions per minute and pushing the ship through the syrupy smooth water at 22 knots per hour. The *Titanic* was not slowing down for anything even though one of the last ice warnings that ended up in the coat pocket of one J. Bruce Ismay showed ice dead ahead.

The director hoped for an early arrival in New York, and though later historians would cast doubt on Ismay's urging of the captain to keep the speed up, it is a safe bet that the hubris of a vessel with a 101-ton rudder infected Captain Smith and Ismay alike. Indeed, second-class passenger missionary Sylvia Caldwell told author Walter Lord that at dinner, "Captain Smith and Mr. Ismay were celebrating what they thought would be the shortest time in which a ship had crossed the Atlantic. These things are better left unsaid, but they are true."[22] Another passenger, Elizabeth Lines, traveling in first class, sat down with her daughter in the dining saloon on D Deck. At 1:30 Ismay and Captain Smith sat down at the table next to them.

She heard Ismay and Smith talking about lighting the last boiler. She said she heard Ismay say the *Titanic* had already beat the speed of the *Olympic* on her maiden voyage and at the speed they were going they would get into New York on Tuesday. Then Captain Smith replied, "Well, we did better today than we did yesterday, we made a better run today than we did yesterday. We will make a better run tomorrow. Things are working smoothly, the machinery is bearing the test, the boilers are working well."[23]

As a concession to the incessant ice warnings Captain Smith had adjusted the course of the *Titanic*, moving the mammoth rudder slightly to a course of S 62 W to S 85. He made this adjustment at 5:50 p.m., thirty minutes after the regular turn that would have occurred at 5:20. This put *Titanic* on a course slightly to the west and south, which many of the officers assumed was a reaction to the ice warnings. Even the venerable old Grand Man of the sea had to admit there was something to the new gadget. The two Marconi operators looked like children to Captain Smith, and he had no idea how wireless technology even operated. Maybe it was time to retire.

Meanwhile, Phillips plowed on though the passenger messages in the wireless shack. "Hardly wait to get back. Cable me awfully happy. Love Mutzie."[24] He moved his hand trying to relax his wrist with the cracking of the blue spark reverberating in the insulated room. The powerful Marconi transformer stepped up the current until the arc of the sending key was like a small explosion. Phillips mouthed the letters, transmitting to the distant station when he jumped in his chair as the new signals exploded in his ears. Volume depended on the strength of the sender's set and distance was the key. The closer the ship, the louder the signal. He quickly deciphered the telegraphic Morse code.

"Say old man, we are stopped and surrounded by ice."[25]

It was the same as getting into a car with radio turned up full blast. The *Californian* must have been practically on top of them. The volume of the transmission pierced Phillips's eardrums and, worse, he lost the message he was sending to Cape Race. The long hours, the staying up late to fix the wireless, the pile of passenger messages all combined into wireless rage. Phillips jammed down on the key with such force the table thumped. "Shut up! Shut up! I am busy. I am working Cape Race, and you are jamming me!"[26]

The *Californian* operator fell silent, and Phillips returned to sending the passenger messages. Fourteen-hour days at $30 a month was enough to try any man, but his nerves were beyond shot, and he glanced at the clock. It was almost 11:30. Phillips knew the *Californian* operator would think twice before he did that again. Ten minutes later he was still tapping the telegraph key in the insulated room and didn't feel the slight grinding that rose up from deep below the waterline of the *Titanic*. It was 11:40 p.m.

CHAPTER THREE

The Race Begins

160 Minutes

FOR THE PASSENGERS, THE FIFTH NIGHT OUT WAS LIKE ALL THE REST, an amazing ride on top of a floating hotel. The ship had a pool, a squash court, a Turkish bath, a steam room, massage rooms, palatial dining rooms, a gym, an electric horse, an electric camel, and the kind of service and food found only in the best restaurants. Royalty was on this ship: John Jacob Astor, Benjamin Guggenheim, Isidor Straus, Margaret Brown, and presidential assistant Archibald Butt. The cream of society loved strolling on deck during the day and at night after dinner, looking at the myriad stars that peppered the heavens. And on nights like this, cold and crystal clear, the stars jeweled the night sky like a frost of Christmas lights.

This was the tenor aboard the *Titanic* as dinner was served in the main dining hall, and seventeen-year-old Jack Thayer, traveling with his parents in first class, explored the deck of the world's largest ship. "The *Titanic* was over four city blocks long—882.5 feet long, 92.5 feet wide, and 94 feet deep. She was as tall as an eleven-story building . . . and had accommodations for 3,500 people."[1] Thayer was headed for Princeton and was enthralled with the *Titanic's* size and amenities. He would later write a book about his experiences, but twenty-eight years would pass before he put pen to paper. He then summed up the world of 1912:

Ordinary days, and into them had crept only gradually the telephone, the talking machine, the automobile. The airplane due to have so soon such a stimulating yet devastating effect on civilization . . . the radio as known today, was still in the scientific laboratory. The Marconi Wireless has just come into commercial use, and the Morse code for help was "CQD."[2]

Thayer evokes a world that made sense, writing this in 1940 on the eve of World War II with Pearl Harbor just around the corner:

The conservative morning paper had headlines larger than half an inch in height. Upon reaching the breakfast table, our perusal of the morning paper was slow and deliberate. . . . Nothing was revealed in the morning, the trend of which was not known the night before. . . . These days were peaceful.[3]

He then draws a clear line between the fading Edwardian world and what came after:

It seems to me that the disaster about to occur was the event which not only made the world rub its eyes and awake, but woke it with a start, keeping it moving as a rapidly accelerating pace ever since, with less peace, satisfaction, and happiness. . . . To my mind, the world of today awoke April 15, 1912.[4]

But twenty-eight years before, Jack Thayer was getting cold as he walked the deck:

I have never seen the sea smoother than it was that night; it was like a millpond, and just as innocent looking, as the great ship quietly rippled through it. . . . The wind whistled through the stays, and the blackish smoke poured out of the three forward funnels. . . . It was the kind of night that made one feel glad to be alive.[5]

Onboard the *Titanic*, the temperature had dropped dramatically to about 24 degrees Fahrenheit, and most passengers took shelter in the saloons, smoking rooms, or their cozy staterooms with electric heaters.

Soon, schoolteacher Lawrence Beesley, like many others, was under his covers and reading his book, while Thayer explored the mostly empty decks and then headed for his parents' stateroom. The throb of the engine was the breathing of the ship, and it gave comfort to the passengers, who only found it annoying, however, when lying in their metal bathtub with the vibrations tickling the back of their necks. Beesley, as he read in bed, noted that the speed of the ship seemed to have increased. He cited two reasons as evidence: "As I sat on the sofa undressing, with bare feet on the floor, the jar of the vibration came up from the engines very noticeably and second, that as I sat up in the berth reading, the spring mattress supporting me was vibrating more rapidly than usual."[6] But Beesley thought nothing more of it and settled down into the quietness of the night with his book, reading while the *Titanic* plowed through the North Atlantic.

Lookouts Frederick Fleet and Reginald Lee would have loved to be in a warm bed and reading a book. The cold wind that hit their faces made their eyes water and numbed their cheeks. It was a brilliant night, and they appreciated the stars, but the night could still hide an iceberg, and though there had been many warnings, the ship had not slowed down. No, the word was she was going to set a record on her maiden run to New York with none other than the director of the line on board, Bruce Ismay.

Up on the bridge the officers were working with little knowledge of what was right in front of them. Of the three last ice messages received addressed to Captain Smith, the *Caronia*'s was posted in the officers' chart room, the *Noordam*'s warning message was lost, and the *Baltic*'s ice warning rode around in Ismay's pocket until he gave it back to Captain Smith. Because of the message's fate, many of the officers thought the ice lay north of them, and few "visualized the great berg-studded floe drifting across the ship's path."[7] Even though Fourth Officer Joseph Boxhall saw the ice warnings, he did not bother to read the one ice warning, a chit marked "ice" above the chart room table. Thus, most of the officers and Captain Smith assumed the *Titanic* would reach ice sometime before midnight. Second Officer Lightoller came on duty and never even saw the chit marked "ice." He was soon joined by Captain Smith, who had left a dinner party shortly before nine. The two men stared out into the black night and the captain remarked on how cold it was.

"Yes, it is very cold sir. In fact, it is only one degree above freezing,"[8] Lightoller replied. He went on to say he had told the carpenter to watch his freshwater supply and told the engine room to keep an eye on the steam winches.

Smith nodded.

"There is not much wind."

"No, it is a flat calm, as a matter of fact."

"A flat calm. Yes, quite flat."[9]

Lightoller then remarked that a breeze would make it easier to see the icebergs at night, with the surf splashing white at the base. Captain Smith stayed on the bridge until 9:25 and then turned to leave. "If it becomes at all doubtful, let me know at once. I'll be just inside."[10] The question is, if the captain was concerned about ice, why not slow down? The answer is complicated, but the nub of it is that the *Titanic* was clipping along at 22 knots and there was no reason to slow down in Smith's mind. He believed an iceberg would present itself in plenty of time to veer around it. This goes to the thinking that the *Titanic* was not in an ice field but *was entering* an area where some icebergs or growlers had been sited. High above the Atlantic up on the bridge, it was inconceivable that the ship deemed unsinkable by the press, if not the captain himself, should bash into an iceberg and be in danger of sinking—although it had happened before. The Guion liner *Arizona* in November 1879 was the largest liner of her day and had collided head-on with an iceberg off the banks of Newfoundland. The berg shattered thirty feet of her bow, but the forward bulkhead was not breached, and she managed to reach St. John's. Another collision with an iceberg occurred when the German Lloyd liner *Kronprinz Wilhelm* caved in her bow and part of her starboard side when she scraped a berg in the early darkness of morning. Closer to home, in 1911, the Anchor liner *Columbia* in a dense fog hit an iceberg off of Cape Race that drove back her bow plates ten feet.

But these things only happened to other ships, and, besides, in the competitive era of transatlantic crossings, no captain worth his salt was not cognizant that setting a record crossing the ocean was good for the ship, the company, and the man. No, Captain Smith was going to bed following the standard operating procedure of his time, cross any ice field as quickly as possible. He had been commanding ships for almost forty years, and his word was law. He was "worshiped by crew and passengers alike. They loved everything about him—especially his wonderful combination of firmness and urbanity. It was strikingly evident in the matter of cigars. His daughter later told Walter Lord in *A Night to Remember*, "Cigars were his pleasure. And one was allowed to be in the room only if one was absolutely still, so that the blue cloud over his head never moved."[11] Also, before wireless there was no real way to know where ice lurked on the dark Atlantic, and the captain of a ship had to rely on his experience; intuition; and, to some degree, luck. Wireless now provided information on the locations of icebergs, but this would necessitate slowing down as one crossed into an area where known ice was reported. To act decisively on a telegraphic ice warning was not standard, and this thinking had not yet crossed the grand old men of the sea, who were masters in their domain. In the case of the *Titanic*, this thinking would cost the ship mightily,

At 10 p.m. First Officer Murdoch took over from Second Officer Lightoller. Murdoch commented immediately on the weather. "It's pretty cold." "Yes, it's freezing,"[12] replied Lightoller, and remarked the ship was reaching the reported ice. Then he headed for his warm bunk. The men high up in the crow's nest were freezing, however, and, worse, they could not find the binoculars that should have been up in the crow's nest. So, Fleet and Lee were squinting, rubbing their eyes, and wiping away the tears that froze on their cheeks. The Atlantic reminded lookout Fleet of a flat tabletop of polished marble. He wiped his eyes again and peered into the black sea. Fleet and Lee had been staring into the darkness for hours along with four other lookouts.

There were a million stars above in the coal-black night and the mirrored glass of the sea was so calm one could skate on it, but in a strange way this clarity played tricks on their eyes. The stars reflected back from the sea, making it hard to see where the horizon actually ended, and maybe that was just one more reason that contributed to the next moment, when all the dreams and magic of the night vanished as a dark monolith appeared in front of the bow. The black mountain approaching lookouts Frederick Fleet and Reginald Lee high above the world in their crow's nest was a building that dwarfed the *Titanic* and weighed 500,000 tons.

Arctic cold smoked Fleet's breath as he squinted at the approaching mountain of black ice. It might have been a berg that turned over, but it did not matter. It was what everyone had feared, and the *Titanic* was steaming right toward the mountain at 22 knots. Fleet banged the crow's nest bell three times and picked up the phone to the bridge. "What do you see," a clipped British voice asked.

"Iceberg right ahead," Fleet answered.

"Thank you."[13]

First Officer Murdoch on the bridge called for a hard turn to stern and told the engine room to reverse the engines. But this massive ship of 52,310 tons fully loaded was not nimble. She had narrowly escaped a collision with the America Line's SS *New York*. She too was a mountain—882 feet and 92 feet wide and 46,328 tons of gross space. These two mountains were now on a collision course, and Fleet watched as the iceberg towered out of the water at more than 100 feet. There was no way this collision could be avoided, and Fleet and Lee braced themselves for the impact.

It had been all of thirty-seven seconds since Fleet called the bridge, and still the ship had not turned. The mountain glistened now and towered over the forecastle deck as the bow headed for the center like a knife plunging into white cream. Fleet set his teeth and then . . . then . . . the bow began to turn away and the moment of impact did not come . . . yet. The iceberg was now on the starboard side of the ship, and Fleet and Lee turned their heads together, each man feeling as if they had just dodged an exceptionally large and lethal bullet.

Later at the Senate hearing, testimony would center on why the iceberg was not seen earlier. "At 22½ knots, the *Titanic* was moving at a rate of 38 feet a second . . . meaning that the berg had been sited less than 500 yards away."[14] The lack of binoculars in the crow's nest was cited next. A reported haze on the water. A blue berg, which was an iceberg that had turned over and was clear ice. None of these really held up as there was no real haze. In fact, the night was absurdly clear; the binoculars were discounted, as they would allow the lookouts to see farther but still not make out something in the darkness. The iceberg was clearly white, as reported by many passengers and crew. No, the best that could be said was the *Titanic* was moving too fast for conditions, and after that it was her fate to meet the skyscraper waiting in the darkness.

Then we turn to Officer Murdoch's evasive actions. He had the ship turn "hard a starboard" and at the same time pulled the engine room telegraph to STOP and then REVERSE ENGINES. This sealed the *Titanic*'s fate. If Murdoch had done nothing, the ship would have smashed into the iceberg bow first and damaged the forward compartments, but the bulkheads would have kept the sea at bay. By turning and reversing the engines, the flow of water past the rudder cavitated, and the ship turned just enough to rip open her side. Then came the awful shudder. Mrs. E. D. Appleton in first class would later say it reminded her of someone tearing a long piece of calico. Four stewards chatting in the first-class dining room felt the shudder as the silver rattled on the tables set for the morning. Lady Cosmo Duff-Gordon, who woke with a start, compared it to someone drawing a long finger down the side of the ship. James B. McCough, a Gimbels employee, had a much ruder awakening when chunks of ice fell in through his open porthole. Quartermaster George Thomas Rowe standing on the bridge noticed tiny shards of ice hovering around the lights and then the iceberg slipped by like a passing train and the shocked Rowe watched it recede back into the darkness.

Most of the *Titanic*'s passengers had turned in, warm under their White Star blankets with their electric cabin heaters glowing red while the great ship traversed the frigid North Atlantic. They did not notice the great shudder or the cease of the thrum of the reciprocating steam engines. Only a few did. Elizabeth Shutes, traveling in first class as a governess, was in bed. When the *Titanic* hit the iceberg, it startled her out of her bunk, but she went back to bed and attempted to go to sleep.

Suddenly a strange quivering ran under me, apparently the whole length of the ship. Startled by the very strangeness of the shivering motion, I sprang to the floor. With too perfect a trust in that mighty vessel, I again lay down. Someone knocked at my door, and the voice of a friend said, "Come quickly to my cabin, an iceberg has just passed my window; I know we have just struck one."[15]

Major Arthur Godfrey Peuchen in his stateroom thought a heavy wave had struck the ship, and Mrs. J. Stuart White had just turned out her light when she felt like the ship had just rolled over a thousand marbles. Managing director of the White Star Line J. Bruce Ismay woke suddenly in his suite and felt that the ship had struck something. He threw a suit over his pajamas, slipped on his carpet slippers and headed for the bridge. A young couple, Mr. and Mrs. George Harder, on their honeymoon, felt the ship quiver and then a "sort of rumbling scraping noise"[16] along the side of the ship.

But Murdoch on the bridge knew exactly what had happened. The RMS *Titanic* had just struck an iceberg at 11:40 p.m. He had done all he could. He had immediately raked the engine handle to STOP and had Quartermaster Robert Hichens turn the wheel hard a starboard and had already pushed the button for the state-of-the-art system sealing the watertight doors. This technology, more than anything else, hung the "unsinkable" moniker around *Titanic's* neck. Captain Smith, the bearded, patriarchal, sixty-two-year-old veteran of White Star Line, who had commanded *Titanic's* sister ship *Olympic* among others in his long career, had been in his cabin when the iceberg ripped along the starboard plates of the ship. He felt the impact and came charging out of his cabin just off the wheelhouse.

"Mr. Murdoch, what was that?"

"An iceberg sir. I hard a starboarded and reversed the engine, and I was going to hard a port around it, but she was too close, I couldn't do anymore."

"Close the emergency doors."

"The doors are already closed."[17]

The measures had been taken. *Titanic* should have been fine, wounded, but the logic of the watertight bulkheads would be the remedy for any breach of the hull. Surely there was damage, but the sea would be sealed off from the other areas of the ship and the damage contained. But

Second Engineer James Hesketh knew better down in the bowels of the ship. The red lights and bells of the bulkhead doors of boiler room compartment four had just begun to sound when he was talking to Fireman Fred Barnett. At that moment, the starboard wall of the ship seemed to collapse, and the Atlantic Ocean swirled into the ship in an avalanche of foaming green water. Barnett and Hesketh leaped through the closing door as it slammed down behind.

The ripping of calico that Mrs. Appleton would compare the sound to was more like a can opener ripping through steel plate with rivets popping out like heated walnuts. The iceberg had struck *Titanic* well below the waterline and then proceeded to tear a long, five-compartment gash that was now a three-hundred-foot vent sucking in water at the rate of four hundred tons per minute. Barnett and Hesketh tumbled into compartment five, which had a gash two feet long beyond the sealed door. This was where the can opener stopped, but seawater was pouring in through the twenty-four-inch hole. If this had been in compartment number four, *Titanic* would not have sunk. But *Titanic's* design had a fatal flaw—the watertight bulkheads did not go all way up through the ship to the ceilings—so as the ship tilted down toward the ocean, the water slopped over the top and filled the next compartment like water flowing up stairs that were inverting.

Captain Smith knew none of this, but Lamp Trimmer Samuel Hemming knew something was not right. He was lying in his bunk when he heard a strange hissing from the bow. He got up and walked to the farthest point of the ship as the hissing increased. He reached forward to an opening over the forepeak locker that housed the anchor chains. Air hit his open hand and moved it up. The air was rushing out at a furious rate like compressed air suddenly unleashed. Trimmer Hemming had no idea that the sea was rushing into *Titanic* so fast it was squeezing out the air in the bow as the ship filled. Captain Smith would have known what all this meant, but he was on the bridge with Officer Murdoch and Fourth Officer Boxhall and had just gone to see if he could still see the iceberg. He could not, and Smith immediately sent Boxhall to inspect the ship. He returned quickly and reported that he went down as far as steerage and the ship seemed sound.

Rushing up the stairs, carpenter J. Hutchinson told a different story. In fact, Captain Smith had just told Boxhall to send the carpenter on a mission to sound the ship, but Hutchinson had already done it and had not had to go far. He looked at the captain with white fear in his eyes. "She's taking on water fast!" Then a mail clerk rushed onto the bridge, gasping out, "The mail hold is filling rapidly!"[18] The cast of characters on the bridge was growing. The director of White Star appeared in his carpet slippers with pajamas still on under his suit. Captain Smith, who technically worked for Bruce Ismay, looked at the mustached man with delicate features and told him the RMS *Titanic*, the pride of the line, had just struck an iceberg. Ismay's eyebrows arched up. "Do you think the ship is seriously damaged?" It was a fearful question. Captain Smith did not miss a beat. "I'm afraid she is."[19]

The *Titanic* royalty at this moment was not the captain or the president of the company, but the man who had built *Titanic*, Thomas Andrews. His technical title was managing director of Harland and Wolff Shipyard, but he was the man who oversaw the building of the ship and was along on the maiden voyage to iron out the kinks. He was a personable man ranking somewhere between Ismay and Smith and the crew, and, so far, he had navigated minor problems in the galley with the stove, the coloring on the promenade decks that had turned out to be too dark, and the minor complaints that are part of any maiden voyage. In fact, he was in his room, A36, pouring over blueprints when the *Titanic* struck the iceberg, and he thought little of the change in ship vibration.

But the captain's summons brought him on the run and he immediately made a quick inspection of the ship, seeing the squash courts with green sea water rippling across the lacquered floor, then the floating letters in the mailroom. He continued making note of the water in the forepeak, No. 1 hold, No. 2 hold, No. 5, No. 6. Water above the keel. He ran into a couple who had eaten at his table in the saloon dining room. "There is no cause for excitement. All of you get what you can in way of clothes and come on deck as soon as you can. She is torn to bits below, but she will not sink if her after bulkheads hold."[20] Andrews was being disingenuous but did not want to start a panic. Also, he was still not sure of the full extent of the damage. Another first-class passenger accosted

Andrews on his inspection tour near the grand staircase. She asked him numerous questions along with stockbroker Harry Anderson, who had seen the "75- to 100-foot iceberg pass by the windows of the smoking room."[21]

Andrews assured them that "*Titanic* could break in three separate and distinct parts, and that each part would stay afloat indefinitely."[22] Then he set off, and what he saw next showed that he had painted a very inaccurate picture of the ship's plight. The seawater was in the two-story mailroom, and luggage was floating around in the first-class hold. When he reached the boiler rooms his worst fears were confirmed. The stokers were trying to douse the fires with the excess steam blowing off outside with the sound of screaming locomotives. Andrews stared at the men working the pumps, but he saw immediately that no less than five of the watertight compartments had been flooded.

Thomas Andrews returned to the bridge and did some quick calculations, deducing that "the facts showed a 300-foot gash with her first five compartments hopelessly flooded."[23] Eighty-four years later an underwater robot hovering over the corpse of the *Titanic* would confirm he was not far wrong, and by using sonar it was found that the ship had six slits of various sizes across 230 feet. "There was a small trace wound near the prow, followed by punctures of about 5 feet, then 6, then 16, then one further aft and slightly lower of 33 and finally, the one visible to submersibles today, a gash of 45 feet."[24]

Andrews told Smith immediately, "Well three have gone (fully flooded) already captain."[25] Captain Smith then left Andrews while he tried to work out how much time the ship had left. The ship's designer thought as much as a third of the ship had been damaged and the ship's "second, third, and fourth compartments were flooding uncontrollably."[26] Murdoch's efforts to turn the ship had saved the first compartment, but the can-opener effect of the iceberg raking the cold steel open had already filled the forepeak tank, which could hold 190 tons. This weight was already plunging the bow down like a lead weight pulling down a canoe.

It was in boiler room 6 that Andrews found the knowledge he didn't want. The water filling the first five compartments was a death sentence.

He knew as nobody else did that the *Titanic* was mortally wounded. She could not float with five compartments flooded, and it was now a mathematical certainty; she would sink. The question was how long could the ship remain afloat? The bulkheads only went up to E Deck. Once the bow sank low enough, the water would overflow into the sixth compartment, then the seventh. . . . Captain Smith was having trouble getting his head around this. Andrews had to assure him several times that there could be no mistake; this beautiful ship would be down at the bottom of the Atlantic probably within ninety minutes, maybe two hours. Andrews was a little off. It would take the RMS *Titanic* 160 minutes to slip beneath the surface. It was only a few years before that Smith had given an interview and remarked on the ship *Adriatic*, "I cannot imagine any condition which would cause a ship to founder. . . . Modern shipbuilding has gone beyond that."[27]

But modern shipbuilding had not anticipated the perfect storm of a ship torn open to the point where the sea could overwhelm the watertight bulkheads with the same certainty as plunging a glass into a tub of water and then tilting it until it filled. Twenty-five minutes after the impact, Captain Smith ordered Chief Officer Wilde to uncover the lifeboats and ordered First Officer Murdoch to muster the passengers; he then sent Fourth Officer Boxhall to wake up Second Officer Lightoller and Third Officer Pittman.

Andrews then began a one-man mission to rouse the ship and yet not instill panic. He walked through the corridors and encouraged the stewards to get the passengers into life jackets and to set an example by wearing them themselves. He suggested to stewardess Annie Robinson, who was working in the empty A Deck staterooms, to "put your lifejacket on and walk about and let the passengers see you." Robinson protested, saying, "It looks rather mean." Andrews lost his composure for a moment and snapped, "No put it on." "He then paused, 'Well if you value your life, put it on.'"[28] The ship's designer was certainly fighting his own sense of panic at this point, and when another stewardess, Mary Sloan, saw him, she "read in his face all I wanted to know."[29]

Captain Smith had left the bridge and quietly walked to the wireless room, the insulated Silent Room on the port side, which was a good twenty yards from the bridge. There were essentially three small rooms. The operator's cabin had the Marconi transmitter and the emergency transmitter along with pneumatic tubes for messages and several clocks for different time zones. Then there was the sleeping quarters along with a wash basin and cupboard, and finally the Silent Room that had the main transmitter, which was connected to a T-type antenna and switching gear. Marconi had sold his systems to lighthouses and ships in the early days to compete with the land-based cable system. By 1912, most of the ships of the North Atlantic carried Marconi sets with relay stations along the coast, and Marconi operators trained at his Wireless Telegraph and Signal Company.

Captain Smith entered the warm room that smelled of heated electrical coils and saw First Operator Jack Phillips standing and Second Operator Harold Bride hard at work with passenger messages. Bride had just signed on. Born in Nunhead, London, England, in 1890, the youngest of five children, after primary school Bride decided he wanted to become wireless operator. He completed training for the Marconi Company (the new name) in July 1911 and was assigned on several ships, including the SS *Haverford* and, more famously, the *Lusitania*. In 1912, he joined the *Titanic* crew as a junior wireless operator assistant.

Phillips had trained with the Marconi Company as well. He was older than Bride, born on April 11, 1887, in Farncombe, Surrey. He had just celebrated his twenty-fifth birthday on the *Titanic* three days before. Jack went to a private school on Hare Lane, then St. John Street's School, and began working at the Godalming post office, where he learned telegraphy. He started with the Marconi Company in March 1906 in Seaforth and graduated five months later in August. He landed on the White Star's ship *Teutonic*, then a series of ships, ending with the *Lusitania* and *Mauretania*. In May 1908, he worked at the Marconi station in Clifden, Ireland, until 1911, when he was assigned to the *Adriatic*, and then in 1912, the *Oceanic* before becoming the senior wireless officer on the *Titanic*.

Phillips and Bride had no idea the ship had struck anything when Bride took over. In fact, he wasn't due to take over until 2 a.m., but he knew Phillips was tired, so he had relieved him a little before midnight and started making his way through the deep well of inbox messages. *Titanic* had been outfitted with the most powerful transmitter available, but the earlier trouble had them using a less powerful method of transmitting. Their bunks were behind a green curtain, and Phillips was just about through the curtain separating their personal space from the Silent Room when Captain Smith stuck his head in the door.

"We've struck an iceberg, and I'm having an inspection made to see what it has done to us. You better get ready to send out a call for assistance, but don't send it until I tell you."[30]

Then he disappeared, leaving Harold Bride and Jack Phillips staring at each other. Phillips took back over as Bride handed him the headphones. Smith already knew the condition of the ship, and it might have been a moment of denial, for he spun around on the deck and reentered the wireless room. Captain Smith had some paper in his hand with their latest position and spoke the words that spelled their true situation. "Send that call for assistance."

Phillips stared at the captain and asked if he should use the regulation distress call, CQD.

"Yes, at once."[31] Smith handed him the paper with the longitude and latitude of their position.

Phillips paused, looking at the paper, then positioned his right hand on the table, his wrist loose, pressing with his fingers. If he had rolled a quarter across the table, he would have seen it veer off to the right and down. The ship was already beginning to list. At 12:15 a.m. on April 15, he tapped out the universal distress code, "CQD; CQ—All Stations. D—Distress," followed by MGY, the call letters for the *Titanic*. He did it six times, shooting out the electromagnetic waves that 2,229 people's lives now depended on. Out into the black space over the Atlantic the CQD bursts emanated like great invisible smoke signals traveling in the night air.

In 1906, the International Radiotelegraphic Convention had agreed on three dots, three dashes, three dots in Morse code as the international distress signal, or SOS. The Marconi Company's distress signal, CQD was still widely used, and SOS was not universally recognized, so many did not send it. Phillips's key cracked lighting, shooting out the same message over and over. Every bit of shipbuilding technology was now subservient to the radio waves being shot off into space from the four-wire antenna strung high over the funnels blasting off excess steam from the boilers like locomotives gone berserk. "CQD . . . CQD . . . MGY . . . 41.46 N, 50.14 W . . . CQD . . . MGY . . ."[32]

The race to save the RMS *Titanic* had just been set in motion.

CHAPTER FOUR

Californian

155 Minutes

MARCONI OPERATOR CYRIL FURSTON EVANS YAWNED AND STARED AT the key just below his right hand. He tapped slowly, the crack of the electricity jumping the contacts rifling around the insulated walls. He finished the message and heard the ticking clock on the wall in the silence and he could feel the engines, but other than that it was the dead silence of the North Atlantic on a calm night. The cargo liner *Californian* was nowhere near as large as the *Titanic* at 6,223 tons, and it was typical for the smaller ships to only have one Marconi man. This put Evans on a grueling schedule of eighteen hours, which began at 5 a.m. and ended at 11 p.m. He usually took a midday nap after lunch so he could be at his set when atmospheric conditions favored wireless transmission.

Evans liked to chat late at night with other operators when traffic had died down. Being a Marconi operator was a lonely business as they had little contact with the crew and technically worked for the Marconi Company. Evans was only twenty-four and what he did was not understandable to most of the men on the ship. Not unlike early computer programmers, Evans was regarded as a bit socially inept. But he had befriended a few of the officers. Third Officer Charles Groves liked to come and chat at night and was very interested in learning Morse. But other than Groves, Evans was left on his own, rising from his bed and sitting down at the set before the sun came up and going back to bed under a night sky. He heard little in the insulated room and only ventured on deck for meals and to take ice warnings up to the bridge. The first eight days had proved uneventful as the *Californian* made her way toward Boston from Liverpool, and wireless traffic had been light.

But on April 14, there had been a marked uptick in ice warnings from other ships that had cited icebergs in the area. It had been an unusually warm winter in the Arctic and large amounts of ice had drifted down from the Greenland Glacier along with the ice from the Arctic icecap that sailed down on the Labrador current. The temperature had dropped as the day began and the first ice warning came in at 9 a.m. from the Cunard liner *Caronia*, reporting bergs and growlers at 42 N from 49 to 51 W. Even though the report was not addressed to the *Californian*, Evans, like many operators, had listened in and then taken it up to the bridge. Then, twenty minutes before midday another ice warning came in from the Dutch liner *Noordam* at about the same position. Evans took that one to the bridge and returned to pick up a warning from the German ship *Amerika*, saying she had seen two bergs at 41.27 N, 50 W.

Then the *Californian* saw some ice at 5 p.m. Groves had just relieved Chief Officer Stewart. Stewart was thirty-five and usually took his dinner with Captain Lord. "He was a competent reliable seaman who had considerable experience on the North Atlantic run."[1] Dining with his captain was a privilege and was the one bit of socializing Stanley Lord did with his officers. Captain Lord was the lord of his ship. He brooked no insolence, and his standards were exacting. He stood well over six feet tall, had a piercing stare and an aquiline nose, and leaned slightly forward when he spoke. The crew regarded him as cold and aloof, and he preferred it that way. Born on September 13, 1877, in Bolton, Lancaster, he was the baby of a family of six boys. His parents had plans for him to go into business, but Stanley, like many young men of the time, wanted to go to sea, with the idea of being an officer in the British Merchant Marine. He persuaded his parents and at thirteen he shipped out, apprenticing with the Liverpool shipping firm J. B. Walmsley Company. Then he joined the *Naiad* out of Liverpool headed for South America. Lord was determined and focused and quickly completed his studies for his "mate certificates," which allowed him to begin his journey toward command of his own ship. After seven years with sailing ships, he joined the Pacific Steam

Navigation Company of Liverpool. By twenty-four he had earned his master's and extra master's tickets. Lord was on his way to commanding a large transatlantic liner and applied for a position with the White Star liner in 1901. He felt insulted when White Star said they would only take him as a third or fourth officer, so he joined the Leyland Line.

Six years later, he had his first posting as captain, and Lord quickly established himself as a no-nonsense, strict commander of the ship.

> *Lord was noticeably aloof and autocratic, not only with his crewman, but with his officers as well. Though not overtly a bully, he was an intimidating presence, and, not surprisingly, his crew generally feared him more than they respected him, in particular dreading his wrath should some job be performed to anything less than his exacting standards, for he had a particular sharp tongue.*[2]

If Captain Smith was avuncular, then Captain Lord was the opposite: remote, distant, unapproachable. The only fractious irritant to his absolute command was the Marconi operator Evans and his wireless. In the past, the isolation on the high seas made the word of the captain next to God and there was absolutely no one who superseded Captain Lord on the *Californian*. Now this squawky box of dots and dashes elbowed its way into his domain. The absolute power of the captain had been diminished by the intrusion of wireless communication and "created a notable abatement in a captain's authority . . . but tradition still imbued him with an aura of near infallibility."[3] Captains still had the ability to conduct marriages, issue death certificates, and arrest and imprison crew or passengers. Lord, like many captains of his time, regarded wireless as something between a toy and a nuisance.

In 1906, he took command of the Leyland Line's *Antillean*, and in 1907 he married and had a son in 1908. All was in line now for Captain Stanley Lord as he moved swiftly from the *William Cliff* to the Leyland Line's *Louisiana* and then finally the *Californian* in 1912. The first crossing was from Liverpool to New Orleans, where she was loaded with cotton, then a stop in New York before a run to Le Havre, France. The crossing was rough, and the bales of cotton were damaged. But then she turned around and on April 5 left for Boston.

After Lord and Chief Officer Stewart left the bridge, Third Officer Groves made note of the icebergs five miles to the south and took it to Evans in the wireless room to let other ships know. Groves was twenty-four and already had his second mate's ticket. He was well educated and confident in his command. Second Officer Stone was also twenty-four and had recently married. He was strangely insecure and worried about losing his position with the Leyland Line but saw the firm commanding exterior of Lord as something to emulate. He might have even regarded Stanley Lord as a father figure.

Cyril Evans tapped out the ice warning at 6:30 p.m. and sent it to the *Antillian* and gave his position as 42.3 N, 49.9 W. Evans listened to the traffic, hearing the *Titanic* sending passenger messages to Cape Race for relay to New York. One hour later, he sent the ice warning to the *Titanic* to let her know of "three large bergs five miles to southward of us"[4] at 42.3 N, 49.9 W. Then another message from the *Mesaba* came in. "Lat 42 N to 41.25 N, Longitude 40 W to 50.30 W saw much heavy pack ice and great number large ice bergs, also field ice."[5]

Then the traffic died down again and he listened to the ships in the area contacting each other and the passenger traffic from *Titanic* for Cape Race. The Newfoundland station had just come in range for the *Californian*, and Evans listened to the first-class passengers on the *Titanic* using the new novelty called wireless. "NO SEASICKNESS. ALL WELL. NOTIFY ALL INTERESTED. POKER BUSINESS GOOD. AL."[6] Meanwhile, Groves took over the watch from Stewart. Stewart told Groves of the ice warnings and waited for fifteen minutes while Groves's eyes acclimated to the darkness. The bridge was kept dark, with the compass light playing under the eyes of both men. Captain Lord walked up and joined the men on the bridge and admonished Groves to "keep a sharp lookout for ice."[7]

Stewart then left. The night was brilliantly cold with the stars reflecting off the still water. The temperature had dipped to 24 degrees Fahrenheit and Groves stared into the black sea ahead as the *Californian* plodded along at 11 knots. Just after 10 p.m. the Third Officer saw some white patches in the water directly in front of the ship. He turned to Lord and said he thought there were some porpoises in the water. Captain Lord squinted and knew instantly the white patches were not sea creatures. He took a deliberate step to the bridge telegraph and rang for FULL SPEED ASTERN. The propellors came to a stop and then cut the icy water in reverse as the ship shuddered and came to a stop. The porpoises were growlers and small bergs, the precursor to a larger ice field. Captain Lord stared into the ice field, "the white patches turned into flat patches of field ice, which soon surrounded the vessel completely. There was no telling how far it stretched or how thick it was, but Captain Lord had never been in ice before, and he was taking no chances."[8]

Lord's approach to an ice field laden with steel-cutting bergs was to wait it out until daylight. A second masthead was lit to show the ship had stopped and warning other ships the *Californian* was not under way. The captain then instructed Groves to keep an eye out for ice and went below to the chart room to catch some sleep on a settee. The silent steamship drifted and turned slowly in the millpond sea. There was no sound at all on the North Atlantic. An Irish voice drifted up from the stokehold ventilator. It was an amazingly clear night, and Groves had never seen the Atlantic so flat. He sipped his tea up on the bridge of the *Californian*. The thrum of the engines had ceased, and there was little to do but stare out into the crystal night of pinwheel stars. The six-thousand-ton ship only carried forty-seven passengers, and Groves knew they would go no farther tonight because of the danger of hitting a berg. He stared into the darkness, looking for icebergs, but something else had caught his eye. It was a ship that had appeared on the coal-black horizon as a glittering carnival of lights.

It was 11:15 p.m., and Groves watched the steady progression of what appeared to be a liner of impressive size. The deck lights lit the night like a Christmas tree, and he counted three decks of illumination. She was big all right, and Groves decided to tell Captain Lord. The ship should know that there was ice in the area. He went below to the chart room and informed the captain, who suggested they signal the ship by Morse lamp. Groves returned to the bridge to set up the Morse lamp when he saw the position of the ship had changed. The ship had stopped, and some of her lights seemed to have gone out. This was not unusual, as liners would douse lights to encourage passengers to go to bed. Groves watched as more lights went out and the ship had turned and was now bow on.

Captain Lord had been watching the ship for some time from his porthole and didn't see the same ship that Groves was seeing. To Lord, the ship looked the same size as the *Californian*. He had stopped into the wireless shack at 11:15 and asked Evans if there were any ships in the area. "Only the *Titanic*,"[9] Evans replied. Lord told Evans then to warn the *Titanic* about the ice and that they had just stopped. Captain Lord then returned to the bridge and looked at the ship though his binoculars. Third Officer Groves suggested it was a large passenger liner. Lord shook his head and dismissed the idea. "That doesn't look like a passenger steamer." Groves frowned. "It is sir. When she stopped, she put most of her lights out—I suppose they have been put out for the night."[10]

Then Groves estimated the distance at less than ten miles. Lord grunted and told him he was going back to the chart room and he was to be informed "if any other ships were spotted, the other ship changed bearing, or anything else unusual occurred."[11] He then left Groves on the bridge alone. In the wireless room Evans had raised the *Titanic* and contacted his old friend Jack Phillips. "Say old man, we are surrounded by ice and stopped—"[12] He didn't use call letters, and he didn't use the letters designating that the message was for the captain. The powerful 5 kW set on the *Titanic* nearly blew off Phillips's ears. "Shut up shut up you are jamming me. I am working Cape Race!"[13]

It was so loud to Evans that it sounded like the *Titanic* was on top of them. He knew they must be very close. So, he listened for a while, but it had been a very long day and he was tired. At 11:30 p.m. he turned off the wireless set and put on his pajamas and climbed into his bunk with a book. The small reading lamp and the absolute silence of the sea at night lulled him, and Evans soon fell fast asleep. When Phillips angrily responded to the very loud intrusion, "Keep out! Shut up! I am working with Cape Race," he had no idea that *Titanic* was sinking, and that he would need everyone within the range of his wireless. The *Californian* was essentially a ship asleep, with only Groves on the bridge, watching the strange ship. He raised the binoculars to his eyes again and stared for a long moment. Groves thought it might be the refraction of the light over distance, but the ship now seemed to be tilting toward the sea.

Thirty-five minutes after Evans had climbed into bed, Phillips sent out his distress calls. The electromagnetic waves fanned out over the open sea, bounced off the ionosphere, and then continued on as far as North America. Phillips kept up the steady stream of dots and dashes, shooting out his distress code CQD MGY, and the first antenna it hit was just ten miles to the west where it ran down from the wire strung up along the mast and then was snuffed out in the cold dead wireless set of the *Californian*. The waves continued to pass over the conducting wire, but they were not transformed by the transformer into the necessary bursts of electricity that were then converted into the letters of human intelligence. They just went on past the ship lying dead in the water turning slowly around with her mast and running lights flashing in the distance. Captain Lord and Marconi operator Cyril Evans slept, while ten miles away all hell was breaking loose.

Chapter Five

An Undoubted Tilt Downward

150 Minutes

Schoolteacher Lawrence Beesley was under his White Star blanket. He had noticed that the "spring mattress supporting me was vibrating more rapidly than usual."[1] Beesley continued reading when he felt an "extra heave of the engines,"[2] and then the vibration that had been so reassuring in his bed was gone. He stood up and slipped on a dressing gown over his pajamas, went out into the corridor, and asked a steward what had happened. "I don't know sir . . . but I don't suppose it is anything much."[3] Beesley had heard of ships throwing propellors, and even though he had only thrown a dressing gown on, he headed up to the boat deck and opened the vestibule door. The arctic wind cut through his thin covering and he saw only a few people milling about. He looked over the side and saw nothing but the black sea, but then he went into the saloon, where he learned the ship had stuck an iceberg.

A group of men playing cards saw the iceberg, and one of them told the schoolteacher, "I am accustomed to estimating distances, and I put it between eighty and ninety feet."[4] Another player pointed to his glass of whiskey and told an onlooker, "Just run along the deck and see if any ice has come aboard: I would like some for this."[5] Another card player remarked, "I expect the iceberg has scratched off some of her new paint . . . and the captain doesn't like to go on until she is painted up again."[6] Beesley felt reassured and returned to his room and got into bed again. But he heard more people in the corridor. This time he put on his Norfolk jacket with two books in each pocket and headed back up to find out what was going on.

Jack Thayer was still out walking before the *Titanic* reached the iceberg and remarked on the stillness of the night. "It was the kind of a night that made one feel glad to be alive."[7] Thayer then visited his parents in their stateroom before going into the adjoining room, calling out "good night." He opened his port window and put on his pajamas and wound his watch to 11:45 p.m. "There was the steady rhythmic pulsation of the engines and screws, the feel and hearing of which becomes second nature to one, at sea. It was a fine night for sleeping."[8] He was standing by his bed when he felt a slight sway. "I immediately realized that the ship had veered to port as though she had been gently pushed. If I had a brimful glass of water in my hands not a drop would have been spilled."[9] Then the engines stopped, and he could hear the breeze whistling though the porthole and then people running in the corridor. The engines started again, but to Thayer they sounded different, "as though they were tired."[10]

Then the engines stopped again. "The sudden quiet was startling and disturbing. Like the subdued quiet in a sleeping car, at a stop, after a continuous run."[11] He threw on a coat over his pajamas and put on his slippers, then went into his parents' room. They too had felt something, and he said he was "going up on deck to see the fun."[12] His father said he would follow after he dressed. When Thayer reached the boat deck the cold was shocking. Like Beesley, he found a few people on deck and some ice on the foredeck. His father joined him, and they learned from a crewman that the ship had struck an iceberg. Then Thayer noticed something more ominous. "The ship took on a very slight list to starboard . . . about fifteen minutes after the collision, she developed a list to port and was distinctly down by the head."[13] Beesley also had noticed an "undoubted tilt downwards from the stern to the bow . . . as I went downstairs, a confirmation of this tilting forward came in something unusual about the stairs, a curious sense of being out of balance and not being able to put one's feet down in the right place."[14]

There was no general alarm. There was no horn or announcement over a loudspeaker or phones buzzing or flashing lights, sirens. This was 1912, and the printed word was still the main form of communication.

Having wireless signals that jumped from the antenna fifty feet overhead was analogous to the early days of the cell phone when few had the technology. So that left the stewards to do the heavy lifting of informing the passengers with the half knowledge supplied to them by the ship's officers. The door to the cooks' quarters swung open with the light blinding the assistant baker, Charles Burgess. Steward Dodd looked at the men, his face wan and tight. "Get up lads, we're sinking!"[15] Dodd continued on, the Paul Revere of disaster, heading for the waiters' quarters where a saloon steward, William Moss, had been trying to rouse the men. Dodd lost no time and shouted, "Get every man up! Don't let a man stay here!"[16] The two stewards rolled on, the responsibility essentially falling to service people to explain to twenty-two hundred people that the ship they are sleeping in, dining on, having sex in was now going to sink in less than two hours into the ice-strewn Atlantic.

Except they couldn't say that. Panic would be immediate. So, the gentle coaxing began of moving first-class passengers to the freezing air of the outside decks. The stewards and other service people themselves had moments of panic as they prepared themselves for a role they never envisioned.

Walter Bedford wore his white baker's coat, pants, didn't stop to put on his underdrawers. Steward Ray took more time; he wasn't worried—nevertheless he found himself putting on his shore suit. Steward Witter, already dressed, opened his trunk and filled his pockets with cigarettes . . . picked up the caul from his first child, which he always carried with him, . . . then joined the crowd of men swarming out into the working alleyway and up toward the boat stations.[17]

The lamp trimmer, Samuel Hemming, who had noticed the hissing in the forepeak, had gone back to bed when the light from the open door broke in and a man shouted, "If I were you, I'd turn out. She's making water one-two-three and the racket court is getting filled up."[18] A boatswain then stuck his head in. "Turn out, you fellows. You haven't half an hour to live. That is from Mr. Andrews. Keep it to yourself and let no one know."[19] You did not want to shock people even though time was of the essence and literally getting up and out could be the difference between life and death. But people had never seen a ship as large as the *Titanic*, and they had no idea what sort of horror awaited them when the ship would break apart and sink. The age of leisure, of a measured pace, was colliding against the looming industrial churn of megadeath.

So, it was not surprising that a bridge game was in full swing in the first-class smoking room, even as whiskey slopped over the edge of glasses as a subtle tilt pulled the ship forward from thousands of tons of inrushing seawater. Lieutenant Stefansson had just picked up his cards and sipped his hot lemonade when a ship's officer blared into the cigar-laden room, "Men, get on your life belts, there's trouble ahead."[20] The same followed in the staterooms. Husbands who had headed out to see what the trouble was returned ashen and grim faced. Mrs. Washington Dodge looked up as the door opened and her husband Dr. Dodge walked in and sat on the edge of the bed. "Ruth, the accident is rather a serious one; you'd better come on deck at once."[21] Farther down in the ship, Mrs. Lucien Smith had the same rude awakening when the light snapped on and her husband began lying to her, "We are in the north and have struck an iceberg. It does not amount to anything but will probably delay us a day getting into New York. However, as a matter of form, the captain has ordered all ladies on deck."[22]

These men knew instinctively or definitely that the ship was sinking and had to couch it in terms palatable to the nineteenth-century view of feminine ability to withstand bad news. Same with the children. When eight-year-old Marshall Drew was awakened by his aunt, he was told that he had go on deck. Children and adults alike were stunned by the news that a ship deemed unsinkable was now in real danger of sinking. The stewards' style of letting people know went from the third- and second-class shouted announcements to gentle knocks on the doors of staterooms. Second-class chief steward John Hardy threw open the doors to twenty-four cabins with a full-throated shout, "Everybody on deck with life belts on, at once!"[23] while in first class Steward Alfred Crawford had to coax an old gentleman into his life jacket and then tie his shoes. "In C-89, Steward Andrew Cunningham helped William T. Stead into his life belt, while the great editor mildly complained that it was a lot of nonsense. In B-84, Steward Henry Samuel Etches worked like a solicitous tailor, fitting Benjamin Guggenheim for his life belt. 'This will hurt,' protested the mining and smelting king."[24]

The steward finally took the life jacket off and made the adjustments, then had to convince Guggenheim he needed a sweater and pulled it over his head. This was still a time when the upper class had personal valets who assisted in dressing. Sometimes people wouldn't answer their doors at all, and stewards had to move on. One door was jammed, and some people broke it down eliciting a threat from a steward to have them all reported for damaging property. It was a quarter after twelve and no one really knew what was going on except for the general exodus to the outer decks where people were standing around in the bitter cold. But there was that strange tilt. The great ship had noticeably gone down toward the sea, and when people walked, they had the sensation of banging against the corridor walls or, worse, going down stairs that seemed to be falling before them. But there was a subtle panic behind the façade, and some people did not stop to dress, fear getting them out of their cabins and onto the upper deck.

Some threw fur coats over nightgowns, pants over pajamas; others did take their time. Mrs. Lucien Smith wanted to go back to her cabin for jewelry, but her husband objected and said it was best at this point not to bother with "trifles." Some men took a revolver with them, others thousands of dollars in cash, some left hundreds of thousands of dollars in bonds and preferred stock on their desks. The halls filled with people still not sure why they were leaving the comfortable warmth of their rooms for this stinging cold, and yet there was the strange listing, the rumors of the ship being mortally wounded by the iceberg, and then the assurance this was nothing more than an elaborate safety drill. No one was talking as they moved in various states of dress and undress.

Mr. Robert Daniel, the Philadelphia banker, had on only woolen pajamas. Mrs. Tyrell Cavendish wore a wrapper and Mr. Cavendish's overcoat . . . Mrs. John C. Hogeboom a fur coat over her nightgown . . . Mrs. Ada Clark just a nightgown. Mrs. Washington Dodge didn't bother to put on stockings under her high-button shoes, which flopped open because she didn't stop to button them. Mrs. Astor looked right out of a bandbox in an attractive light dress.[25]

And so, it went, with each class separating on the decks, first class in the center, second a little aft, and third class in the stern or shoved up by the well deck in the bow. Nobody was moving toward the lifeboats yet, and many believed at any moment they would be heading back to their cabins. But inside the insulated wireless room Jack Phillips and Harold Bride could have told them they would not be going back to their cabins or staterooms ever again. They were either going into a lifeboat, of which there were only twenty, or they were going into the North Atlantic. The people standing around did not know that their lives now depended on two young men firing off electromagnetic waves into the unforgiving darkness. *CQD CQD require assistance position 41.46 N, 50.14 W. struck iceberg Titanic.* Phillips tapped and tapped and tapped, his finger rapidly playing the key with blue sparks flashing off the walls. Long fuzzy buzzing followed by short bursts. *CQD CQD CQD. . . .* The lives of some 2,200 people now depended on the skill of two young men using new technology no one really understood. *CQD CQD require assistance . . . 41.46 N 50.14 W . . . CQD . . .*

CHAPTER SIX

Marconi Wireless Station, Cape Race

140 Minutes

DISTRESS SIGNALS FROM THE *TITANIC* WERE BEING PICKED UP ALONG the Atlantic coast. The ship was the first to carry a 240-cycle rotary spark transmitter and the signal shooting out into the frigid night was strong. Fourteen-year-old Jimmy Myrick was alone in the Cape Race Station and he was freezing. The coal stove had burned down, and the temperature had dropped into the low twenties. Freezing rain hammered the window outside as he went across the room to put more coal in the stove. He went back and sat down. The equipment glowed in the dim light of the single overhead bulb. He was a wireless apprentice and loved listening to the chatter, decoding the short bursts of Marconi code. Out in the Atlantic, ships were sending messages that reached the station; he could decode them, and that was simply amazing. Myrick imagined the sparks shooting across the cold black ocean, reaching the antenna strung up over the roof, and then coming down to the set that chattered in his ear.

Lately they were messages from the *Titanic* on her maiden voyage, which was incredible. The largest, most famous ship in the world was sending messages nonstop to their station. They were now the closest land-based Marconi receiving port. The two men who ran the station, J. C. R. Goodwin and Walter Gray, had just left for needed repairs, and that left Myrick alone to listen in with strict instructions to write down any messages that came through. The messages of greetings to people back in New York had died down, and the assumption was that the apprentice could handle whatever followed next.

The Cape Race Station was three hundred miles west of the *Titanic*. The iceberg that stopped the ship was grounded on the southern edge of the Grand Banks of Newfoundland, and Cape Ray had become a reporting station for the icebergs that drifted by like a fleet of whaling ships. Eventually the bergs would end up in the graveyard of all icebergs, in the Gulf Stream. The Cape Race Station, along with another station 140 miles west at Cape Ray, had been broadcasting iceberg warnings all day. The stations were studies in isolation with only a lighthouse and the station and no accessible roads. The only access was by sea, and this ensured the occupants ate their meals at the lighthouse and slept in the Marconi stations. These were the crude outposts of communication on the night of April 14, 1912, battling winds, rain, snow, and the monotony of long hours, where the only company was the eerie long buzz of the wireless.

Jack Phillips had been sending messages all day to his friend at Cape Race, Walter Gray. The two had attended the Marconi school together and were in close contact passing on the *Titanic* passengers' messages. It was during a break that Gray left Myrick by himself in the station while he and Goodwin worked on the diesel motor outside. The set was quiet, and Myrick fingered the Morse key and leaned back in the chair. He was for the moment the only ears to the world of the SS *Titanic*, the most luxurious ocean liner ever built and said to be virtually unsinkable. He had heard stories of ships before wireless being isolated on the sea with only flares or rockets to let people know they were in distress. Many times the tale was told by the few survivors fished out of an unforgiving ocean. Mostly the ships took their secrets with them to the bottom.

The wind blew against the window. April in Newfoundland was miserable, and the damp cold seeped into the room. Myrick blew on his fingers, staring at the quiet set. Short bursts of electrical impulses suddenly buzzed in his ears and he began deciphering the Morse code. Tingles crawled up his neck. It was the *Titanic*. He knew this immediately from the call letters MGY. It was probably another passenger communication, but the young boy felt like a giant had just tapped him on the shoulder.

He quickly wrote down the letters coming in over and over, and he felt hot fingers rush up over his spine and pinpoints of sweat ripple across his brow. He checked the letters again and again, but the code kept coming. *CQD*. It was the universal distress call, and it was coming from the largest, fastest, ship in the world.

CQD CQD . . . Titanic Position 41.44 N, 50.24 W. Require imme-diate assistance. Come at once. We struck an iceberg. Sinking[1]—*12.17 a.m., 15 April 1912*

Myrick tore off the headset and ran to find Goodwin, who was coming back in the door. Goodwin listened to the boy and ran out to find the station head, Gray, who was now checking the generators that powered the station. Goodwin almost shouted, "My God, Gray, the *Titanic* has struck an iceberg!"[2] Walter Gray rushed back in and listened for a moment and then began relaying the information to the other land-based stations. Years later, people would say it was amazing that a four-teen-year-old boy, Jimmy Myrick, was the first person on land to hear Jack Phillips's CQD and to understand that the RMS *Titanic* was indeed sinking into the icy Atlantic. Fifty years later the story would finally be told to a newspaper by a relative, after Myrick had been sworn to secrecy by Gray. No one must know that a young boy was the first person in the outside world to receive the news that the RMS *Titanic* had struck an iceberg and was sinking.

Another teenager, sixteen-year-old Charlie Ellsworth, was sitting in the Cape Ray Station 250 miles west of the *Titanic* and 140 miles from Cape Race. Both of the Newfoundland stations were studies in isolation, and the shift from midnight to 4 a.m. was the loneliest. The men who ran the station were in bed and Ellsworth had orders to never start the "ker-osine engine to operate the high-power transmitter, the racket of which would awaken the other operators sleeping in the station."[3] More than fifty years later, his wife told a newspaper what happened next.

Just after starting his midnight watch he wound up the clock works in the magnetic detector and settled down for a quiet, peaceful watch.

Outside the Cape Ray fog horn was broadcasting its mournful reverberations. He took off his headphones to load more coal into the stove feeling a chill that might have been premonitory but was more likely traceable to the cold drizzling foggy weather. My husband was halfway across the room when he heard that sound—it came floating out like a beautiful magic note, emitting from the headphones on the operating table. "CQD CQD . . . MGY WE HAVE STRUCK AN ICEBERG IN LATITUDE 41.46 N AND LONGITUDE 50.14 W APPROXIMATE POSTION AND THE SHIP IS LISTING BADLY PLEASE STAND BY."[4]

Ellsworth tore off the headphones to wake up the sleeping operators, who began transmitting the signals out to the world. Out into the darkness the signals bounced, hitting the ionosphere, bouncing back down, glancing off the ocean, ricocheting to the east, coming out of the darkness to the glittering palaces of New York City. High on top of the Wanamaker department store a lone antenna grabbed one of the signals like an outfielder making a leaping catch. The signal shot down through the wires into another cold wireless room where a young man was leaning back in his chair, his eyes closing slowly.

David Sarnoff was an adventurer who had returned from the ice fields in 1910 and was always looking for the newest thing. The newest thing was Marconi wireless, and Sarnoff sat with the headphones pinching his ears, his arms crossed, sleep nibbling at his consciousness. He was cold and a little bored. John Wanamaker was always looking for the newest trend or invention for his department store and decided to equip his stores in New York and Philadelphia with powerful Marconi radio stations. The station in Manhattan was on the very top of the store, a wireless shack perched on the roof with an antenna facing the Atlantic Ocean. Sarnoff, the dark-haired young man who had applied for and secured the job of operator, found himself alone in the station the night of April 14, 1912.

Fog had enshrouded New York City and a persistent drizzle beat on the roof. The novelty of wireless was one that Wanamaker thought customers would appreciate, and with a nose for the trends of a quickly evolving urban culture, he knew wireless telegraphy would allow him to expand his reach into new markets. But right now, all was quiet as Sarnoff leaned back in his chair and adjusted the tuner. He leaned his head back and closed his eyes. There was no traffic. It was 11:25 p.m. Eastern Standard Time.

The earphones crackled and buzzed as Sarnoff jumped forward, shaking his head, and he touched the radio tuner. The signal was very faint, the buzzing irregular, but he could make it out and began translating the Morse code. He followed the words, mouthing them as he went along. It was from the ship *Olympic* fourteen hundred miles out to sea—but no, it was from the *Titanic*. Sarnoff listened to the repeating signals coming through the static while the rain beat on the small radio shack roof. He felt a chill run down his spine and then center in his bowels. The static overwhelmed the signal again, but it didn't matter, he had the message, and it left him feeling cold and faintly nauseous with the implications of the words. "Titanic to M.K.C. We are in collision with berg. Sinking head down. 41.46 N, 50.14 W. Come as soon as possible."[5] Sarnoff stared at the words and then grabbed for the telephone.

CHAPTER SEVEN

A Thousand Railway Engines

135 Minutes

SECOND OFFICER CHARLES LIGHTOLLER WAS IN HIS BUNK AND "JUST about in the land of nod, when I felt a sudden vibrating jar run through the ship . . . a distinct and unpleasant break in the monotony of her motion."[1] Lightoller leaped out of his bunk and ran onto the deck in his pajamas. He peered over the port side and then ran across to the starboard side. Peering over the sides he saw nothing, but he knew they had struck something. He retreated from the cold and went back to his bunk. Lightoller had not been called for and he had no phone, so the best he could do was lie in his bed and wait to be called. It was far better than roving around the ship trying to ascertain what had happened. He was off watch and expected to be in his room, so that is where they would come to find him. The door swung open at 12:10 p.m. and Fourth Officer Joseph Boxhall spoke quietly.

"We've hit an iceberg."

"I know we have struck something," Lightoller said getting up and pulling on his coat.

"The water is up to F Deck in the mail room."[2]

Lightoller nodded and finished dressing. He was the perfect second officer for the *Titanic*. He didn't get rattled, and he moved with efficiency. He headed up to the boat deck where the sixteen lifeboats were being uncovered. These, along with four canvas collapsible Engelhardt lifeboats, were all that the *Titanic* had to rescue more than two thousand people.

The boats, fully loaded, could only hold 1,178. But no matter, the *Titanic* was not going to sink. The ship was unsinkable. There had been no boat drills and no boat assignments. They simply were not needed, nonetheless the crew began to unfurl the coverings. Yes, the *Titanic* was short lifeboats, but it could have been worse. In 1912 the Board of Trade required only enough boats for 962 people or 27 percent of the 3,547 people she was certified to carry. But because the *Titanic* was a luxury ship, the White Star Line had added space for an extra 216 people.

Lightoller could now hear the roar of the funnels blasting out the excess steam from the twenty-nine boilers. "The instant the engines stopped, the steam started roaring off at all eight exhausts, kicking up a row that would have dwarfed the row of a thousand railway engines thundering through a culvert."[3] The result of this was that speech was rendered useless between officers, crew, and passengers. The best Lightoller could do as he took over the lifeboats on the port side was use hand signals. "A tap on the shoulder and an indication with the hand, dark though it was, was quite sufficient to set the men about the different jobs, clearing away the boat covers, hauling tight the falls and coiling them on deck, clear and ready for lowering."[4]

It is hard to imagine the terror in the passengers as they reached the deck and saw the lifeboats being unfurled and lowered all under the apocalyptic roar of the steam being released. They could not ask questions; they could not talk to one another, "the appalling din only adding to their anxiety in a situation already terrifying enough in all conscience."[5] All Lightoller and his crew could do was "give them a cheery smile of encouragement"[6] as the bosun's mate indicated with his hand his job of clearing away had been completed and Lightoller made a diving motion with his hand to commence swinging out the boats. Now to Lightoller and others it was clear the *Titanic* was listing and the "ship was seriously damaged and taking a lot of water."[7]

Little knots of men swarmed over each boat, taking off the canvas covers, clearing the masts and useless paraphernalia, putting in lanterns and tins of biscuits. Other men stood at the davits, fitting in cranks and uncoiling the lines. One by one the cranks were turned.

The davits creaked, the pulleys squealed, and the boats slowly swung out free of the ship. Next a few feet of line were paid out, so that each boat would lie flush with the boat deck . . . or, in some cases, flush with Promenade A directly below.[8]

Lightoller was frustrated that Chief Officer Wilde had not given him permission to load the boats yet. Wilde was holding off because he had no order from Captain Smith. One could not blame Wilde. The mood was still that of a slightly strange holiday where in the lit-up gymnasium just off the boat deck none other than the Astors were sitting on some mechanical horses. Astor was showing his wife what the inside of the life jacket looked like by slicing it open. Other passengers joked and waited, some smoking cigars, others still holding drinks from the first-class smoking room. The question was out there, why go into one of those little boats when in all likelihood they would just have to return?

The problem was that Captain Smith had not given the order to load the lifeboats. Was there some sort of denial going on with Smith? He had given an interview a few years before saying he had not had any real near mishaps at sea in all his thirty years of command. And now he was watching the pinnacle of his command, the maiden voyage of the *Titanic* degenerate into a catastrophe that had no precedent. Surely Thomas Andrews could not be correct in his assessment that the ship was mortally wounded. But he was. Captain Smith no longer had to look at the ship's clinometer to see the degrees she was listing. The ship had a downward slope like a board being pushed down in front by an invisible hand.

Lightoller saw the main deck was almost level with the water and found the captain on the boat deck. He drew him into a corner and cupping his hands both over his mouth and the captain's ear, shouted, "Hadn't we better get the women and children into the boats, sir?"[9] Captain Smith nodded. But it was Lightoller's reason for wanting to get the boats in the water that is stunning. He wrote later in a book of his experience: "I could see a steamer's lights a couple of miles away on our port bow. If I could get the women and children into the boats, they would be perfectly safe in that smooth sea until this ship picked them up."[10]

The lights were a couple of miles "away on our port bow."[11] This goes against the mythology of the ship going down by the head as the band played on, with only the heroic actions of the doomed to guide people. No one could hear the band, first of all. The concussive shriek of compressed steam overwhelmed all. But the motif of the great catastrophe being out in the lonely frozen night of the Atlantic is shattered as well. There was a ship in visual sight that an experienced officer like Lightoller estimated was *only a few miles away*. This changed the *Titanic* from the scenario of certain doom that has enveloped all histories of the ship to the very real possibility that help was just a few miles away. Of course, the question was, who was this ship?

The steam release suddenly stopped, and the ensuing silence was eerie. A passenger asked Lightoller if he thought it was serious. Second Officer Lightoller assured them it was "more of a precaution to get the boats in the water . . . and I pointed out the lights on the port bow which they could see as well."[12] This was not some distant light to Lightoller. This was a ship close enough to tell the passengers that if worse came to worse that ship would come rescue them. Writing years later, Lightoller pegged the ship as the *Californian*. Also, it comes to light that Lightoller believed he might fill the boats from the gangway on the lower deck. This would be at least one reason the lifeboats were loaded with shockingly few people. He gave the order to a bosun's mate to take six hands and open the port lower deck gangway. The men were never seen again and Lightoller later speculated: "They gave their lives endeavoring to carry out this order, probably they were trapped in the alley way by a rush of water."[13]

Charles Lightoller then returned and lowered Boat 4 until it was level with Deck A and announced that it was for women and children only. He then moved on to Boat 6 and "with one foot in No. 6 and one on deck, Lightoller now called for women and children."[14] There were some protests by people who pointed out they would have to come back as John Astor pointed out, "We are safer here than in that little boat."[15] Grudgingly the women and children began to board Boat 6. Mrs. J. Stuart White climbed into another boat, and someone called out, "When you get back, you'll need a pass. You can't get back on tomorrow morning without a pass."[16]

It was here, supposedly, that the band assembled by bandmaster Wallace Hartley began to play ragtime, adding to the unreality of the moment. "Everything had been done to give the *Titanic* the best band on the Atlantic. The White Star Line even raided the Cunard *Mauretania* for Hartley. Pianist Theodore Brailey and cellist Roger Bricoux were easily wooed from *Carpathia*."[17] The Edwardian age was now asserting itself one last time with music to load lifeboats by. This would be held up as a WASPy ideal of heroism as the band literally played on while the ship was sinking. To a sane person who knew the dire situation the *Titanic* was in with her too few lifeboats and well over half the ship looking at certain death in the icy Atlantic, it must have seen surreal if not outright lunacy to have this classical band serenading passengers looking at one-hundred-foot drops to the sea below, which was the equivalent of staring down a high cliff into darkness.

Or the band didn't play at all. Mrs. May Futrelle, a passenger on deck, recalled fifty years later what was happening while the boats were loaded. "An orchestra did not stand on the deck of the sinking ship and play. . . . They had been playing in the lounge," she recalled. "They couldn't have played on the deck since it was tipped at such a precarious angle. Besides, it was so bitterly cold that night that the strings on the violins would have snapped had a bow been drawn across them outside."[18] But the mythology of the *Titanic* must have accompaniment, and it should be classical music to be saved or die by.

Margaret Freulichen was twenty-two and this trip was a reward for a good report card, but she and her mother had spent the last three days in their cabins with seasickness. The night of the sinking she felt better and ate some supper, but then felt sick again. She returned to her stateroom but heard her mother cry out in the adjoining room, "A collision!"[19] She and her parents then went up on deck where she felt sick again and returned to her cabin. Her father asked an English stewardess if there was danger, "Yes sir, there is. Take your life jackets and go to the boat deck."[20] Margaret managed to come out of her room, still feeling sick. "I remember the steward was standing in the hall. He was pale as a sheet and said to me, 'Don't be afraid, Miss. Remember all the compartments. The ship can't sink.'"[21]

Margaret and her parents rushed up to the boat deck, where two fat sailors lifted her mother into a lifeboat and then hoisted her in alongside. "My father started instinctively to follow us . . . the boat was not at all full, but the sailors said, 'Ladies first sir.' Father moved back and called out, 'Auf wiedersehen.'"[22] Then the boat began to be lowered, but one end stuck and "we were hanging crazily there so they pulled us back up. . . . I started screaming for my father and tried to get back on deck. . . . A British officer said, 'Let the men in.'"[23] Margaret's father joined his wife and daughter, and the boat was lowered away once again. This intact family was one of the lucky few.

Managing director Bruce Ismay knew what was happening and he was beginning to lose his mind. He had a hard time believing the words that came out of Captain Smith's mouth. The *Titanic* would be underwater in a matter of hours, and not only did he have concern for the passengers, but also self-preservation had gripped him like a dark, slimy force coming up from that frigid water. He did not want to die in the North Atlantic, and he was already eying a spot on the lifeboats for himself, but now he was in frenzy and was rushing among the lifeboats proclaiming, "There's no time to lose!"[24]

Ismay still had his pajamas on under his pants and overcoat and was wearing carpet slippers. Third Officer Pittman worked on Boat 5 and saw no reason he should listen to this strange man who was shouting at him to load the women and children. He bluntly told Ismay he was waiting for the commander's orders. He then realized who the bespectacled man was and saw Captain Smith, walking to him and asking if it was the director of the company who had just given him an order and should he follow it. Captain Smith nodded. So, Pittman ordered the ladies and children in, and Boat 5 began its journey down to the sea. But Ismay could not restrain himself. "Lower away! Lower away!" he shouted, waving his arms. "If you'll get the hell out of the way," Fifth Officer Lowe shouted at him while trying to work the davits, "I'll be able to do something. You want me to lower away quickly? You'll have me drown the whole lot!"[25]

Ismay stared at him then turned and walked away. Still others were reluctant to get in the boat. When there were no more women and children, some couples were allowed and a few single men. The problem was the boats were not getting filled anywhere near capacity.

> *First Officer Murdoch, in charge of the starboard side, was having the same trouble filling Boat 7. Silent movie star Dorothy Gibson jumped in, followed by her mother. Then they persuaded their bridge companions of the evening. . . . Others trickled in until there were 19 or 20 in the boat. Murdoch felt he could wait no longer. At 12:45 he waved No. 7—the first boat down.*[26]

Clearly no one believed yet that this floating city might dip below the waves of the icy Atlantic. But Captain Smith believed it as he headed once again for the wireless room to find out if anyone had responded to the CQD. He did not really believe in the power of wireless before, but he now knew it was their only hope. Smith, though a veteran of the seas, was in a state of semi-shock. He had just been given an updated position for the *Titanic* and he wanted to get that into the hands of Phillips and Bride as soon as possible.

At some point, Smith passed a group of stewards milling around the bridge and told them, "Well boys, do your best for the women and children, and look out for yourselves."[27] It was hardly a confidence-inspiring moment, but the truth was that Smith himself had lost confidence in his ship and maybe himself. One can only imagine that with the plethora of ice warnings and the full-speed-ahead attitude he and Ismay had maintained that he had to be feeling some guilt. He still had in his pocket the ice warning from the *Baltic* he had retrieved from Ismay, who had shown it to some passengers as evidence of the state-of-the-art communications the *Titanic* had onboard. At least on this score, Bruce Ismay had it right.

CHAPTER EIGHT

A Clap of Thunder

130 Minutes

AT FIRST NOT EVEN MARCONI UNDERSTOOD THE TECHNOLOGY THAT Jack Phillips and Harold Bride were using to try to save more than two thousand people in the middle of the Atlantic. When he was trying to get a cathode ray to react to electromagnetic waves, Marconi stumbled on the idea of making a large antenna. And when he tried to push waves across the Atlantic the first time, he had no idea if it was possible at all. In Poldhu, England, he had made a large, cone-shaped antenna in the belief it would help the Hertzian waves coalesce. But he had no idea if a Hertzian wave even existed. All he could deduce was that the taller the antenna, the farther the wave traveled. He still had no idea why a ship could transmit over a thousand miles at night and by day maybe four hundred miles at best. No one knew anything about an ionosphere that bounced the waves back to Earth.

No one knew anything really about Hertzian waves, what they looked like, how they traveled, how they acted, and what they could pass through. So, when it came time to try to transmit across the Atlantic Ocean, all Marconi could deduce was that he would need monstrously large antennas and incredible voltage to send the wave on its way. The problem was, the increased power brought all sorts of unforeseen problems. At the station in Poldhu, even the gutters of nearby houses became conduits for current and shot out blue sparks. George Kemp, who was assisting in construction and design wrote, "We had an electric phenomenon—it was like a terrific clap of thunder over the top of the masts (antenna) when every stay sparked to earth in spite of the insulated breaks. This caused the horses to stampede and the men to leave the ten-acre enclosure in great haste."[1] When the key was pressed to send the high voltage spark across the electrodes, it sounded like a clap of thunder.

At the receiving station in Cape Cod, the strange electrical eccentricities persisted, and when the transmitting was tested the "spark lit the sky with such intensity that it was visible and audible four miles down the beach."[2] Even the station cooks hanging the wash felt mild electric shocks as the wet clothes became conduits for the ambient current. Then a storm blew down the giant mast antenna at Cape Cod before any tests could be run and Marconi had to take a different approach. The giant antenna was out. He would use a floating antenna that could reach for the sky. He would use a kite.

So, the bet that Captain Smith, Jack Phillips, and Harold Bride were making on this new technology was grand. Never before had so many lives depended on wireless technology than at this moment. It was the grand test. Think of the *Titanic* as a pole in the center of a great space and from her pole out went the Hertzian waves into the night sky and where they would land no one knew but they hoped they would slather cold steel antennas strung up on other ships, that wireless operators had not turned in for the night, and the ships would not be too far to render assistance. Out went the Hertzian soldiers and they bounced off the *Californian* and made it to Jimmy Myrick sitting in the cold station at Cape Race in Newfoundland. They even rumbled toward David Sarnoff on top of Wanamaker's department store. But it was ships that the *Titanic* needed. There was no Coast Guard. There was no helicopter to pluck people from a sinking ship. There were only other ships that could turn about and make for a stricken liner whose passengers only had 160 minutes to live.

Wireless had proved itself before and made operator Jack Binns the first Marconi star. The Lloyd Italian liner SS *Florida* had collided with the 15,000-ton, 570-foot White Star liner *Republic*, a single-funnel twin-screw steamer, off the coast of Nantucket, Massachusetts, on January 23, 1909. The bow of the *Florida* caved in, and the *Republic* started taking water amidships with her engine room and boilers quickly getting submerged. The *Republic's* captain, Inman Selby, got the passengers out onto the deck, and Binns fired off the new distress signal, CQD. The wireless room had been crushed in the collision, with three of the four walls gone and the power gone as well. Binns went to a storeroom looking for batteries and had to dive underwater to find them. He then quickly rigged his Marconi set and began sending out weak but audible signals. The Coast Guard cutter *Gresham* and White Star's *Baltic* soon came to the rescue and saved the lives of sixteen hundred passengers. Wireless telegraphy had saved the day, and Jack Binns was given a ticker-tape parade in New York and became a household name with songs, films, and a play produced called *Via Wireless*. Binns had become a star, but onboard wireless was an even bigger star.

And so, the *Titanic's* CQD went out to the ships of varying distances, nationalities, and speeds, with the captains who would make all the difference. The *La Provence* received the first CQD from the *Titanic*. Then the *Mount Temple*, the *Frankfurt*, the *Virginian*, the *Baltic*, the *Mauretania*, the *Parisian*, the *Cincinnati*, the *Prinz Frederick*, the *Olympic*, and the *Carpathia*. And each ship that received the signal passed it farther on to another ship and another and then back to Cape Race and on to New York. The night lit up suddenly with electromagnetic waves as the urgency of the moment took hold. The *Titanic* was sinking quickly, and there was not a moment to lose. Within an hour, the entire world was aware of the drama unfolding on the North Atlantic. This had never happened before. Wireless telegraphy was the CNN of the moment, and it was on the spot. One of the untold consequences of Phillips sending out his distress call was real-time reporting of what was happening. The *Titanic* was an international ship registered as part of a British company, the White Star Line, but it was owned by an American international banker, J. P. Morgan. Add to that the cream of society and a personal advisor to President Taft on board and one can understand the reaction of a fourteen-year-old boy in a remote wireless station and a twenty-one-year-old young man sitting on top of a department store in New York City.

And now every amateur wireless operator in his home along the Atlantic seaboard was picking up the signals binging through the cold night air. They were many, and they would have a role to play in spreading the news by relaying it on. It was the first "share" predating the internet by seventy years. It was simply the biggest story of the age, and it was now unfolding in real time. The response of each ship, land station, and individual would determine how many lived or died on the *Titanic*. But the clock was ticking.

CHAPTER NINE

Bursts and Crashes of Controlled Static

128 Minutes

TWENTY LIFEBOATS ARE ALL THE *TITANIC* HAD, WITH A CAPACITY FOR 1,178 of the 2,229 passengers aboard. Four of the twenty were collapsible Engelhardts with canvas sides for emergency situations. These were on top of the officers' quarters around the bottom of the ship's two funnels. In theory, these collapsible boats were more accessible than the wooden lifeboats. They were designated with letters A through D, while the lifeboats were numbered. The wooden lifeboats were painted white and hung from davits, except for Boats 1 and 2, which had already been swung out when the *Titanic* left Southampton in the event a passenger fell overboard.

The odd-numbered boats were on the starboard side and the even numbers on the port side. The lifeboats were in horizontal groups of four, and since they had to be swung out for people to get in, the question became whether to load from the boat deck or the promenade. It would be easier on the passengers to board from the promenade, but Captain Smith had forgotten that most of the promenade was shuttered. On his previous command, the *Olympic*, the promenade was open, but on the *Titanic* the passengers had to board from the boat deck, stepping over a space with a ninety-foot drop to the ocean below.

Movie star Dorothy Gibson was in the first boat to be lowered, which was number 7. Her escorts, William Sloper and Frederick Seward, accompanied her. They had heard about the squash courts flooding and believed, as most did not at this point, that the *Titanic* would sink. Sloper later wrote,

> *I felt as certain as anyone could feel that we had come to the end and that many, if not all, would soon be gone. All of the people who were there in this companionway at this time passed out quietly onto the deck where the lifeboats were. . . . The covers had been taken off the lifeboats, and they were quickly swung off from the davits and lowered to the level of the deck. From this deck, we were, if I remember correctly, somewhere about eighty feet above the water, and to leave a well-lighted ship that at the time seemed to have listed slightly, and step into a small boat that might plunge down into the darkness below, or, if it reached the sea safely, be capsized by the water, was a question which made some people hold back.*[1]

The rule of women and children boarding first was not followed as strictly by First Officer Murdoch on the starboard side versus Second Officer Lightoller, who would permit only women and children to enter on the port side of the *Titanic*. Boat 7 was way under capacity and Murdoch let anyone enter the boat who wanted to. Sloper, Gibson, and Seward were in the boat along with a Dutch car salesman, and a Manhattan socialite, Margaret Hays, got in carrying her Pomeranian, Bebe. Sloper would later write that J. J. Astor and his young pregnant wife were offered a seat on Boat 7, but he demurred, pulling his wife away from the boat. Astor, who would perish, probably didn't believe the *Titanic* would really sink and thought it more dangerous to be in a small boat on the Atlantic. He went back into the warmth of the first-class gymnasium, where he famously entertained his wife by cutting open a life jacket. Whatever the reason, it was a fatal mistake, as his next encounter would be with Lightoller, who would make no exception for him to accompany his wife.

Some steamer rugs were thrown in for warmth and Boat 7 was lowered away, with Sloper "expecting one end of the boat would drop faster than the other and that we should all be thrown out into the sea."[2] The harrowing descent through the freezing night ended with a plop into the placid sea where Sloper began to row, but Gibson immediately began to scream. A drain plug had not been inserted and the boat was taking on water. Her feet were getting soaked with the freezing sloshing sea. The missing drain plug could well have sunk the boat if the passengers had not improvised with a plug of their own. Gibson tore off her stockings and the plug was filled, keeping the ocean at bay. Overall, the lowering of the lifeboats was a haphazard operation at best.

Jack Thayer, his father and mother, and a friend named Long stood around on deck not sure what to do. The steam had started discharging again. "The noise was terrific. The deep vibrating roar of the exhaust steam blowing off through the safety valves was deafening; in addition, they had commenced to send up rockets."[3] Thayer and his family waited while his mother watched the lights of a boat off the port side. His mother was seeing the same lights that Lightoller saw and had directed lifeboats to row toward. "The ship was at most not more than ten or twelve miles away," Thayer wrote later, and concluded it was the *Californian*. "While I did not see the masthead lights of the SS *Californian*, many claim they did see them. . . . Second Officer Lightoller told me he positively saw them."[4]

Thayer and his mother and father then retreated from the cold into a crowded hallway where the stewards passed the word: all women and children to the port side. "We then said goodbye to my mother at the head of stairs on A Deck." Thayer, his father, and Long went to the starboard side to await orders, but "no orders ever came." Thayer observed that "people like ourselves were just standing around, out of the way. The stokers, dining room stewards, and some others of the crew were lined up, waiting for orders. The second- and third-class passengers were pouring up onto the deck."[5]

They decided to go find out if Thayer's mother had made it onto a lifeboat and found her still waiting on the port side. The ship had such a substantial list that the lifeboats were swinging away from the ship, and porters "stretched folded steamer chairs across the space, over which the people were helped into the boats." They then headed down to B Deck where they were separated from Mr. and Mrs. Thayer. "Long and I could not catch up and were entirely separated from them. I never saw my father again."[6] The last sentence is shocking, but this was how it was ending for many people. A separation that seemed somewhat innocuous but was, in fact, for eternity.

The ship's designer Thomas Andrews tried to get people into the boats, but Andrews found that all his talk for the past few days of the *Titanic's* invincibility was haunting him. People simply didn't believe it would sink. Passengers joked they would need their tickets to board in the morning, and "You'll be back in the morning for breakfast"[7] was heard more than once. The Edwardian code was still in play at this point, and Margaret Brown later told the *New York Times*,

> *The whole thing was so formal that it was difficult to realize it was a tragedy. Men and women stood in little groups and talked. Some laughed as the boats went over the side. All the time the band was playing. . . . I can see the men up on deck tucking in the women and smiling. It was a strange night. It all seemed like a play. . . . Men would say, "after you," as they made some woman comfortable and stepped back.*[8]

This fits into the Edwardian mythology of the *Titanic's* sinking that made it so much more palatable for readers, but the terror was still yet to come. The size of the ship was such that there were many who still were unaware that it was sinking. This was because there was no protocol beyond Thomas Andrews, Bruce Ismay, Captain Smith, and the stewards spreading the word in a very matter-of-fact fashion, with Andrews helping a woman into a boat who urged him to get in as well; he demurred, saying, "Women and children first."

Incredibly, there were even members of the crew who did not realize the ship was slowly sinking by the bow into the Atlantic Ocean. Quartermaster George Thomas Rowe walked along the stern. At the very back of the 882-foot ship he had heard nothing since the collision with the iceberg and assumed that the damage was minor. He stared down from his lonely watch and then squinted. There was a white boat floating by in the darkness full of people in life jackets. Rowe could scarcely believe his eyes as Boat 7 went by with Sloper rowing along. He immediately went to a phone and rang the bridge and asked if they knew there was a lifeboat in the water. A voice asked who he was and then told him to come at once to the bridge and bring rockets with him. Rowe went down a deck and retrieved a tin box of twelve rockets from a locker and headed for the bridge, realizing only then that the *Titanic* was in danger of sinking.

In the wireless room, Jack Phillips was trying to decipher the mass of static and bursts emanating from the ships and shore stations all transmitting on top of each other. The clean short beeps associated with Morse code transmission were not what Phillips and his other operators heard in 1912. The open spark transmitters sent out "bursts and crashes of controlled static which resembled nothing so much as the interference distant lightning will create on a radio."[9] Add to this that since sending his CQD, the operators were all stepping over each other's signals.

The first replies came in a torrent of static. The German Lloyd steamer *Frankfurt* sent back, "Ok, stand by."[10] Then the Canadian Pacific's *Mount Temple*, the Allan liner *Virginian*, and the Russian tramp *Birma* acknowledged the CQD. The ships were putting about and steaming toward the coordinates Phillips had given them. Captain Smith had come back with new coordinates and asked what call he was sending. "CQD." Harold Bride chimed in. "Send SOS; it's the new call, and it may be your last chance to send it."[11] Phillips laughed and commenced sending the SOS as well. A timeline of transmissions can be seen. This is the beating heart of the beginning rescue operations and shows the frantic repeating quality of Phillips's transmissions. The radio log paints a picture of hope and confusion.

12:15 a.m. *La Provence* receives the *Titanic* distress calls.[12]

12:15 a.m. *Mount Temple* heard the *Titanic* sending CQD. Says require assistance. Gives position. Cannot hear me. Advise my captain his position 41.46 N, 50.24 W.

12:15 a.m. Cape Race hears *Titanic* giving position on CQD 41.44 N, 50.24 W.

12:18 a.m. *Ypiranga* hears CQD from *Titanic*. *Titanic* gives CQD here. Position 41.44 N, 50.24 W. Require assistance (calls about ten times).

12:25 a.m. C.Q.D. call received from *Titanic* by *Carpathia*. *Titanic* said, "Come at once. We have struck a berg. It is a CQD O.M. Position 41.46 N, 50.14 W."

12:25 a.m. Cape Race hears MGY (*Titanic*) give corrected position 41.46 N, 50.14 W. "Require immediate assistance. We have collision with iceberg. Sinking. Can hear nothing for noise of steam." Sent about 15 to 20 times to *Ypiranga*.

12:27 a.m. *Titanic* sends following: "I require assistance immediately. Struck by iceberg in 41.46 N, 50.14 W."

12:30 a.m. *Titanic* gives his position to *Frankfurt* and says, "Tell your captain to come to our help. We are on the ice."

12:30 a.m. *Caronia* sent C.Q. message to M.B.C. (*Baltic*) and C.Q.D. "MGY (*Titanic*) struck iceberg requires immediate assistance."

12:30 a.m. *Mount Temple* hears MGY (*Titanic*) still calling CQD. Our Captain reverses ship. We are about fifty miles off.

12:36 a.m. DKF *Prinz Friedrich Wilhelm* calls MGY *Titanic* and gives position at 12 a.m. 39.47 N, 50.10 W. MGY *Titanic* says, "Are you coming to ours?" DFT *Frankfurt* says, "What is the matter with you?" MGY *Titanic*: "We have collision with iceberg. Sinking. Please tell captain to come." *Frankfurt*: "Ok Will tell."

12:38 a.m. *Mount Temple* hears *Frankfurt* give MGY his position 39.47 N, 52.10 W.

12:45 a.m. *Titanic* calls *Olympic* SOS.[13]

Phillips needed to hook one ship with his bait. He needed one ship that was close enough to make a difference to be able to turnabout and come at once. Phillips's frustration boiled over as he snapped at an obtuse radio operator on the *Frankfurt* to keep out and stop jamming his signal. The problem was that many of the wireless operators had a hard time believing the unsinkable ship was sinking. Phillips knew most of the ships were too far away, but he only needed one nearby to come and rescue passengers and crew before the *Titanic* slipped away.

The unsinkable ship was now leaning forward, and there was no doubt in the mind of Second Officer Lightoller about what was taking place. He stared off the port bow and raised his binoculars. The brilliance of the night made visibility amazing, and what was more amazing was that the lights of a steamer were not more than ten miles away. He did not move, thinking the steamer might be coming toward them. He went for the Morse lamp and began signaling the ship. There was no response. Stronger measures were needed. This ship must understand that they were in dire circumstances and to come at once. Quartermaster Rowe brought the rockets to the bridge and Captain Smith ordered him to fire them every six minutes.

The first rocket shot up from the starboard side of the bridge. Up, up it soared, far above the lacework of masts and rigging. Then with a distant, muffled report it burst, and a shower of bright white stars floated slowly down toward the sea. In the blue-white light Fifth Officer Lowe remembered catching a glimpse of Bruce Ismay's startled face.[14]

This was how Walter Lord described the rockets in *A Night to Remember*, and Bruce Ismay had a right to be startled. Rockets meant one thing: a ship was in distress and, more than that, a ship that was in danger of sinking. The people on deck milling about still had an air of disbelief, even with the rockets bursting overhead. But deep in the bowels of the ship in boiler room number 5 there was no doubt at all. It seemed all was under control with the pumps keeping up with the inflowing water. Lead Fireman Barrett had sent the stokers to the boat stations while he and engineers Harvey and Shepherd worked the pumps. The boiler room was a steam bath from the water used to put out the coal fires and the men moved about in the dim light that hid the manhole Shepherd fell into, breaking his leg. Firemen George Kemish and Barrett lifted him up and took him to a pump room, which was a closed-off space. They went back to working the pumps when the sea burst through the bulkhead between boiler rooms 5 and 6. The wall of water engulfed the men and drowned Harvey as he went to rescue Shepherd. The other men scrambled up the ladder watching the Atlantic Ocean rush in beneath them. The water would continue to rise and eventually would engulf the dynamos powering the ship and Phillips's wireless transmitter that was shooting out Morse signals as he frantically looked for the ship that would save them all.

CHAPTER TEN

Mount Temple

125 Minutes

TWENTY-ONE-YEAR-OLD JOHN DURANT RETURNED TO THE WIRELESS shack on the eleven-thousand-ton steamer *Mount Temple* and yawned. He often worked until one in the morning and, like many wireless operators, he would sleep in after the midday meal so he could catch the night signals when the range of his wireless jumped up. It was 11 p.m., and after delivering the ice warning from the *Corsican* to the bridge, he slipped on the headphones again and listened to the traffic from Cape Race. Captain Moore had been astonished at the ice warning from the *Corsican*. Durant had heard Captain Moore remark as he left the bridge, "I have never in all my experience known the ice to be so far south."[1] The ice was far south and even though a day's journey away, it was directly in the path of the *Mount Temple*. "We are not making the corner where we planned anymore," he said to the officer of the watch. "I am going to the chartroom to redraw."[2]

The *Mount Temple* was heading for New York. She carried 1,466 souls and had none of the grace, size, or beauty of the *Titanic*. She was a cargo ship that had been converted from carrying cattle to carrying people. Her engines only generated about 17 knots at full speed, her wireless at most two hundred miles at night. She was the dark horse of the transatlantic trade, with a stout captain who had just celebrated his fifty-first birthday with his wife, Mary, and his sixteen-year-old daughter. And now he was on the bridge, a veteran of thirty-two years with twenty-seven years on the North Atlantic. Captain Moore was looking at the ice warning brought to him by the wireless operator that told him there was ice dead ahead even though it was a day's sail away. The company policy was very clear to Captain Moore; you did not risk your ship or the passengers' lives by entering an ice field. You avoided it any way you could, and with that in mind he went down to the chart room to alter their course, not knowing he would be less than fifty miles away from the largest ship in the world sinking from a mortal wound inflicted by a iceberg of 1.5 million tons.

Titanic was sending on message after message, and the *Mount Temple* operator, Durant, knew Jack Phillips and thought about chiming in but just listened, keeping an ear for another warning of ice. His set had nowhere near the power of the *Titanic* and the messages Phillips was sending were getting louder. At some point Durant put on his pajamas and was getting ready for bed with the headphones still buzzing on the table. He typically worked until 1 a.m., and the lull of the ship plus the dim incandescent lighting was making him sleepy. He sometimes took a three-hour nap the next afternoon to catch up on his sleep. He did not have the passenger messages that *Titanic* operators had to deal with, and he preferred the later hours when the wireless signals were stronger.

Durant pulled out a book and put his hand behind his head. There was just the buzzing of the wireless that he deciphered while he read, catching the call letters and the gist of the message. The CQD he heard from the *Titanic* at 12:15 literally meant, "*All Ships Attention*," danger that sounded like *buzzzzzz short buzz buzzzzzz*. Durant looked over and stared at the copper headset. The call letters shouted at him. *MGY*. Durant stared at the headphones for a few seconds, dropping his book, whipping back the covers and sliding into the chair with the headphones already on. He tapped back a request for more information and rang the bell for the night steward, glancing at the clock.

Phillips tapped his initial CQD at 12:15 a.m. The signal shot through the darkness, with the long and short buzzes hitting the Cape Race Station in Newfoundland but also getting picked up by three ships: the *Mount Temple*, the *Birma*, and the French liner *La Provence*. Durant listened again with his bare feet on the cold floor of the wireless room, tapping out his ship's call sign and asking for more information. The buzzes erupted in his ear. *Titanic* said it could not understand his incoming message but repeated the CQD with coordinates of their position. Durant quickly scribbled the message down and wrote URGENT across the top and gave it to the night steward to take to Captain Moore and told him to hurry.

Looking at a map of the ships closest to the *Titanic* and in range of Phillips's CQD, then, one would start with the *Californian* less than 10 miles away. And then equally distant were the *Parisian* and the *Mount Temple* at less than 50 away. Then the *Carpathia* at 50 miles, the *Birma* at 70 miles, the *Frankfurt* at 140 miles, the *Baltic* at 200 miles, and, farthest away, the *Olympic* at 400 miles. Captain Moore was asleep in his cabin when the night steward knocked on the door. The captain opened the door, reading the message, mouthing the words, looking at the steward as if for confirmation before shouting up the speaking tube to the second officer to turn the *Mount Temple* around and head back east. He then hurriedly slipped on a pair of pants and an overcoat and sprinted up the stairs toward the bridge.

Captain Moore told his second officer that the *Titanic* had struck ice and was sinking and had him increase the speed of the ship. The captain then descended to the chart room to work out a more specific course and returned to the bridge and had the second officer change direction. He then sent a message to wireless operator Durant to be alert for any more information. There was not a moment to lose, and he sent for the chief engineer, who had been asleep and was still groggy when he reached the bridge. Moore told him to get every stoker out of bed and down to the engine room to fire up the boilers to full steam and divert all the steam to the reciprocating engines. The engineer went down into the ship shouting to the men to turn out. Captain Moore then posted men in the bow to watch for ice and sent a message to the man in the crow's nest to also watch for icebergs and any signals he might see from the *Titanic*.

Captain Moore felt the thrum of the propellors as the old 1901 cargo steamer protested the increased speed. The stokers were now throwing coal into the furnaces as fast as they could, and the cylinders of the engines began to beat in time while the sea foamed white around the bow and stern. Moore then ordered the chief officer to "have all the boats uncovered and swung out. All of them. Then checked and provisioned."[3] He then had the ship's surgeon prepare rooms and the infirmary for a potential emergency. The *Mount Temple* was the first ship to turn around and was one of the first ships to react to Jack Phillips's CQD. She was the closest now in the race to reach the stricken *Titanic*, while the *Californian*, less than ten miles away, slept on.

Chapter Eleven

CQD

120 Minutes

JACK PHILLIPS SENT OUT THE CQD TO ANYONE WHO COULD CATCH IT and possibly help.

> *CQD MGY* [Titanic] *SINKING CANNOT HEAR FOR NOISE OR STEAM*
> *CQD MGY I REQUIRE ASSISTANCE IMMEDIATELY. STRUCK BY ICEBERG IN POSITION 41.46 N, 50.14 W*
> *MGY TO DFT* [Frankfurt] *MY POSTION 41.46 N, 50.14 W. TELL YOUR CAPTAIN TO COME TO OUR HELP. WE ARE ON THE ICE*
> [Caronia *to* Baltic] *MGY STRUCK ICEBERG REQUIRES IMMEDIATE ASSISTANCE*
> *DFT TO MGY. WHAT IS THE MATTER WITH U?*
> *MGY TO DFT. WE HAVE COLLISION WITH ICEBERG. PLEASE TELL CAPTAIN TO COME.*
> *DFT TO MGY. OK WILL TELL. MY POSTION 39.47 N, 52.10*
> *MGY CQD CQD . . .*[1]

The Russian ship *Birma* heard the CQD but few on board knew English. Still, the Russian radio operator heard the CQD and so did his captain. Captain Ludwig Stulping deduced it was a distress call. Who sent it and why he had no idea, but he had the coordinates and turned the ship around. Stulping understood that CQD translated to "Come quick danger." But even in this basic understanding he was wrong. Marconi's code book defined CQ as meaning that what followed was a general signal meant for all parties. The D meant "Distress." The Russian captain turned his ship around and headed east northeast, even though he had no idea who was signaling, why, or what their situation might be. Such was the power of the CQD among mariners. They saw it as a request to take every measure necessary to come to the aid of a ship that was in danger of sinking. The response of the *Birma* shows how these three letters were enough to invoke extraordinary measures on an international scale.

While Phillips and Harold Bride feverishly worked to contact anyone who could come to their rescue, Lawrence Beesley was standing on deck watching the lifeboats being lowered when he heard the first rocket explode.

Suddenly a rush of light from the forward deck, a hissing roar that made us all turn from watching the boats, and a rocket leapt upwards to where the stars blinked and twinkled above us. Up it went, higher and higher, with a sea of faces upturned to watch it, and then an explosion that seemed to split the silent night in two.[2]

Beesley and others knew what the rockets foreboded. "And with a gasping sigh one word escaped the lips of the crowd: Rockets!"[3] Beesley observed some third-class women attempt to enter a lifeboat and were blocked by an officer who told them, "Your boats are down on your own deck."[4] Then Beesley saw a bandsman. "I saw the cellist come round the vestibule corner from the staircase entrance and run down the now deserted starboard deck, his cello trailing behind him, the spike dragging along the floor."[5]

The schoolteacher then went to the railing and watched a boat being lowered away and heard First Officer Murdoch instruct them to row to the gangway. The strategy of loading from below had still not been given up. A voice from below floated up from the darkness, "Any more ladies?" Beesley looked around. The crewman in Boat 13 called up twice more and then noticed Beesley. "Any ladies on your deck?" Beesley replied, "No." The crewman paused. "Then you had better jump."[6] Beesley jumped onto the rail with his feet hanging over. He threw his dressing gown down into the boat and then "fell in the boat near the stern. As I picked myself up, I heard a shout, 'Wait a moment, here are two more ladies,' and they were pushed hurriedly over the side and tumbled into the boat."[7] The crew then shouted to lower away but not before a man and his baby tumbled into the boat with the baby first being handed to a woman in the stern and the man falling in after. These acrobatics in the frigid darkness were being performed while dangling a hundred feet over the North Atlantic in a wooden lifeboat. Lawrence Beesley, by being in the right place at the right time, would be one of the saved. It was repeated over and over, a capricious life-and-death process with no real order. If Beesley had not been by the railing, he might never have made it into a boat.

Many were not so lucky. Women and children went into the boats while husbands, sons, valets, uncles, and grandfathers watched. Seamen grabbed wives, sisters, and mothers and guided them to the too-few lifeboats. Mrs. Charlotte Collyer was being dragged to a boat with her husband calling to her, "Go Lottie! For God's sake be brave and go! I'll get a seat in another boat."[8] This was happening as the unbelievable roar of escaping steam from the boilers still blasted out of the funnels and rockets were bursting overhead. It was apocalyptic and no amount of lying could cover the fact that these were partings of forever. But some refused to go.

A famous story of the *Titanic* was that of the Strauses. Mrs. Isidor Straus, wife of the Macy's department store owner, refused to leave her husband. "I've always stayed with my husband; so why should I leave him now?"[9] They had seen it all and risen from the ashes of the Confederacy to build Macy's into a national and cultural department store and then a stint in Congress; now they were in the afterglow of great success with a trip to Cape Martin and the maiden voyage of the *Titanic* to return to America. Isidor Straus pleaded with his wife to go. She refused, knowing her fate, handing her jewelry to her maid. "Archibald Gracie, Hugh Woolner, and other friends tried in vain to make her go. Then Woolner turned to Mr. Straus. 'I'm sure nobody would object to an old gentleman like you getting in.' Straus refused. 'I will not go before other men,' he proclaimed, and then he and Mrs. Straus sat down on a pair of deck chairs."[10]

While this bit of chivalrous drama was being played out topside among the tuxedoed one percent, down in the bowels of the ship no one would escape. "Men struggled desperately to keep the steam up . . . the lights lit . . . the pumps going. Chief Engineer Bell had all the watertight doors raised after Boiler Room 4—when the water reached here, they could be lowered again."[11] Forty-five ventilation fans were switched off as fires were doused, turning the boiler rooms into sweltering hellish chambers of smoke and steam. Trimmer George Clavell was drawing down fires when he felt the water rise to his knees. He then went up the escape ladder but looked back to make sure he had not left behind any of the other men. He paused then and went back down slogging around in the water almost up to his waist before going up the ladder again.

Up on deck the partings among the first-class passengers continued as the ship slanted toward the sea. Walter Lord described these chivalrous partings in *A Night to Remember*, and one can only assume he assembled these from the survivors themselves.

> *"It's all right little girl," called Dan Marvin to his new bride; "you go, and I'll stay a while." He blew her a kiss as she entered the boat. "I'll see you later," Adolf Dyker smiled as he helped Mrs. Dyker across the gunwale. "Be brave no matter what happens, be brave," Dr. W. T. Minahan told Mrs. Minahan as he stepped back with other men.*[12]

Dr. Minahan would drown, but his body was recovered. His effects give an interesting snapshot of someone who was traveling first class at the time and had time to collect some things from his stateroom before going topside. Recovered from his body were a "pocketknife, papers, gold watch engraved 'Dr. W. E. Minahan,' keys, fountain pen, clinical thermometer, memo book, tie pin, diamond ring, gold cuff link, nickel watch, comb, check book, American Express Check for 380, 1 collar button . . . 14 shillings, nail clippers."[13] He was interred in Green Bay, Wisconsin, but in 1985, trophy hunters broke into his mausoleum and stole his skull. It was later recovered.

Then more scenes from Lord's *A Night to Remember*. Mrs. Tyrell Cavendish says nothing to Mr. Cavendish. "Just a kiss . . . a long look . . . another kiss . . . and he disappeared into the crowd."[14] Sons, even of the first class, are not allowed into boats. Charles Fortune and his son see off Mrs. Fortune and their three daughters with admonishment from Mr. Fortune: "I'll take care of them: We're going in the next boat." A final call comes from one of his sisters to Charles: "Charles, take care of father."[15] Wives plead with their husbands, as Mrs. Walter P. Douglas begs Mr. Douglas to come with her. "No," he says. "I must be a gentleman," before turning away forever. "Try and get off with Major Butt and Mr. Moore," she calls back. "They are big strong fellows and will surely make it." Arguments break out in the open at a time when couples never fought openly, and so Mr. and Mrs. Edgar Meyer go down to their cabin to argue about her leaving, while Arthur Ryerson commands his wife into a boat: "You must obey orders. When they say women and children to the boats, you must go when your turn comes. I will stay here with Jack Thayer. We'll be alright."[16]

These scenes of the first-class Edwardian partings on the deck of the *Titanic* are at the heart of the mythology that pops up in many books. It is as if a play is taking place on top of the sinking ship with all the pathos of bad melodrama. Sons must send on mothers not willing to lose their children. Alexander T. Compton listens to his mother's refusals, then calmly replies, "Don't be foolish mother. You and sister go in the boat—I'll look out for myself."[17] Along the same lines, Mr. and Mrs. Lucien Smith do not go down to a cabin but argue in the open on the deck. They are a young couple and presage a different world where people could disagree in a marriage. Mrs. Smith then implores Captain Smith, standing nearby with his megaphone, to make an exception to the women and children rule and let her husband go. Captain Smith does not skip a beat, exhorting women and children to get into the boats. Finally, Lucien stares at his wife and lies, "I never expected you to obey me, but this is one time you must. It is only a matter of form to have women and children first. This ship is thoroughly equipped, and everyone on here will be saved."[18] Mrs. Smith looks in his eyes and asks if he is telling the truth. He never wavers. "Yes." They kiss, and she steps into a boat as he watches his young bride descend toward the Atlantic, calling out, "Keep your hands in your pockets; it's very cold weather."[19]

It's not that these partings are necessarily untrue, although since Lord's book is not footnoted, we have to take his word. Again, it is more where Lord chooses to put his camera, and these partings must be emphasized to hold up the heroic ideal of the WASP patriarch of the early twentieth century. For how else to justify the carnage? If the partings are not chivalrous and honorable, then they must be something else. Something that lurks in human psychology rooted in the primal urge to survive, which is also true, but not pretty.

The steam continued roaring, the rockets bursting, and still off the port bow were the strange lights on the horizon that everyone assumed was a ship coming toward them. People now returning to staterooms to retrieve a last-minute bauble, a stack of money, stocks, a picture were meeting a wall of green water filling the hallways. Some retreated up the stairs and ran directly into a boat. A maid was nearly locked in her employers' room when a steward locked the door to prevent looting; only her shriek saved her from certain death. Again, this was the civil Edwardian mythologized, the polite part of the *Titanic* sinking that made for great copy. It was the top rung of society waltzing along toward death on top of the hordes of passengers still below, mostly third class. As the boat's angle increased and rockets persisted, and the lifeboats began to paddle away, a very different scene was evolving. Chief Officer Wilde knew what would happen when there were but a few boats for thousands of people facing drowning in the icy Atlantic. He asked Second Officer Lightoller along with the captain and First Officer Murdoch to help him find the firearms. He went to the locker where they were kept and stuck a pistol in Lightoller's hand, remarking, "You may need it."[20]

Charles Lightoller went back to loading the boats and watching the water creeping up an emergency stairwell used by the crew to gauge how fast the boat was sinking. "That cold green water, crawling its ghostly way up the staircase, was a sight that stamped itself indelibly on my memory.

Step by step, it made its way up, covering the electric lights, one after the other, which for a time, shone under the surface with a horribly weird effect."[21] Lightoller was glad the dynamos were still running and the lights still on as he continued to assist women and children into the lifeboats that hung out over the dark space. "Standing with one foot on the deck and one in the boat, the women just held out their right hand, hooking my left arm underneath their arm, and so practically lifted them over the gap between the boat's gunwale and the ship's side, into the boat."[22]

Lightoller now knew the ship was doomed and labored to lower all boats away as quickly as possible before the *Titanic* slipped under the waves. He realized that the ship he could still see in the distance might be a false hope as well. "Why couldn't she hear our wireless calls?" he wrote later. "Why couldn't her officer of the watch or some of her crew see our distress signals with their shower of stars?"[23] He felt worse, as he had sent the boats toward the light. "D'you see her—the ship over there. . . . The breeze should spring up at daybreak, and she will be able to come pick us up from the boats."[24]

It gets worse for those in the lifeboats.

For the Schwarzenbach family, who managed to get off with their family all together, they had a close encounter with the ship. "Not much later those in the lifeboat saw the lights of what proved later to be the freighter *Californian*."[25] The ship was so tantalizingly close "they screamed in unison toward the boat, but it passed on."[26] That the passengers in the Schwarzenbach lifeboat should feel a ship was so close they might hear them is almost unbelievable. But the notion the *Titanic* was all alone in the North Atlantic is obliterated by these sightings, one after another. Lightoller later lamented he didn't have a six-inch gun to wake the ship up with a couple of shells. In truth, that might have been the only thing that would have made Captain Stanley Lord of the *Californian* move his ship.

CHAPTER TWELVE

Carpathia

115 Minutes

HAROLD COTTAM, ALL OF TWENTY-ONE, SAT IN THE WIRELESS ROOM OF the *Carpathia* listening to the traffic out on the North Atlantic. He rubbed his hands together and looked at the electric heater on the wall. The interference produced fuzzy static, and he turned his tuner several times for different bandwidths. If someone picked up a headset in 1912 on the night of April 14, they would hear long static then short beeps and long beeps in succession. To listen in on other people's wireless messages was extremely easy. In the early days of the telephone, in rural areas people were on the same circuit. The party line allowed people to listen in to a neighbor's conversation, and that neighbor had to wait until the line was free to initiate their call. A lot of people found out other people's business this way, and this was exactly the case in wireless radiotelegraphy in 1912.

Cottam was the youngest graduate ever of the Marconi school, at seventeen, and he had to wait until he was twenty-one to get his first posting at sea. He worked for four years at a shore station and then he had his first posting on the White Star Line's *Medic*. After four months, he joined the *Carpathia* under Captain Rostron, who was known as a fair skipper who understood the power of wireless much more than other captains in 1912. Cottam listened to the ice warnings, some operators stepping on top of others. This make it hard for wireless operators do their jobs, but the new technology had no standards. Some ships like the

Titanic and *Olympic* had immensely powerful sets, while others had weak transmitters only capable of a couple hundred miles. "There were a half-dozen types of equipment; two different Morse codes, American and International; no regulations concerning the hours the wireless watch was to be kept."[1]

The volume of work for young Cottam and others was another problem. The frivolous messages by passengers competed with the ice warnings and messages of distress, which were the real function of wireless, at least in the eyes of the operators. And on a smaller ship like the *Carpathia* there was only one operator who had to work fifteen-hour days, taking his meals at his desk and getting little sleep when his days stretched to eighteen hours. The *Carpathia*, while much smaller than *Titanic* at nearly 14,000 tons, had been designed to carry foods in three refrigerated compartments. Her engine was a "pair of ten-cylinder quadruple expansion engines turning two propellers, which gave her a top speed during her sea trials of just over 15 knots."[2] She could only carry second- and third-class passengers, but with the constant immigration to the United States as well as the wealthy Americans who enjoyed the Mediterranean crossings, she was given a retrofit in 1905 and converted to a passenger liner with 100 first-class passengers, 200 second-class cabins, and third-class accommodations of 2,250.

It was mostly the first-class passengers who kept Cottam busy with their messages to New York and other ships. But the ice warnings had been increasing, and he had to balance the passenger messages with trips to the bridge. Compounding all of this was the reality that there was no clear direction on what messages should be taken to the bridge for the captain. If a message was tagged for the captain, then it was obvious, but on this Sunday night Cottam was listening to the unusual number of ice warnings on the northern track of the Atlantic and he had already run a few up to the bridge when the traffic lulled. He breathed in the faint scent of cigar smoke, as the wireless room was above the second-class smoking room. He sat back in the chair and listened to the distant ships trading news on the ice and relaying messages to Cape Race in Newfoundland. At 7 p.m. he heard from Jack Phillips aboard the White Star liner *Titanic*. It was a direct message for *Carpathia*, a ship-to-ship message from one of *Titanic*'s first-class passengers to a Mrs. Marshall on the *Carpathia*.

Cottam had noted *Titanic*'s silence most of the afternoon, and he wondered if there were some problems with its equipment. Phillips was not much for chatting when the night was long and traffic died down. He listened as Phillips relayed messages to the Marconi station in Cape Race, admiring his mastery of Morse code. He would send one message, then wait a minute to see if any other ships needed to send a message, and then he sent another. A few more ice warnings came in during Phillips's pauses, from *Mesaba* and the cargo ship *Californian*.

At 10 p.m., Cottam went to the bridge again with the accumulated ice messages just as Captain Rostron, First Officer Horace Dean, and Second Officer James Bisset were taking over for the second watch. Captain Rostron studied the messages and noted the ice warnings were far north of them and turned to Bisset, "Wonderful thing wireless isn't it?"[3] The man who appreciated the power of wireless was coincidentally known throughout his company as the "electric spark" for "his decisiveness and boundless infectious energy."[4]

Born in 1869 in Astley Bridge near Bolton, Lancashire, the young man wanted to go to sea at an exceedingly early age and joined the cadet school HMS *Conway* in Liverpool at thirteen. He was apprenticed after two years on an iron-hulled clipper ship *Cedric the Saxon*. For six years he sailed all over the world while getting ready for mate's examinations. Rostron was serving as second mate on the *Red Gauntlet* when in a storm off the coast of New Zealand she "toppled over her beams end"[5] (literally lying on her side). The ship recovered and young Arthur Henry Rostron survived. In 1894 he passed his extra master's certificate and joined the Cunard Line. He was given a position on the ocean liner *Umbria*. He worked his way up with a break where he served in the Royal Naval Reserve in 1905 during the Russo–Japanese War. In 1907, he became first officer of the brand-new *Lusitania*, and right before her maiden voyage he received command of the cargo ship the *Brescia*, then *Ivernia*, *Pannonia*, and *Saxonia*. When he took over *Carpathia* on January 18, 1912, he was heading toward the zenith of his career, respected widely as a fair captain who ran a tight and efficient ship. Captain Rostron was a man who did not drink or smoke or swear and who turned to God in times of crisis. Piety and humility were words used in describing him by his officers and crew alike.

The *Carpathia* traditionally carried a large contingent of immigrants in her third-class sections on trips between Fiume and New York, and then on her return she would carry first- and second-class passengers bound for holiday in Europe. When she pulled away from Cunard's Pier 54 on April 11, 1912, and headed out of New York harbor, she was carrying 125 first-class passengers, 65 in second class, and 550 third-class passengers in a space designed for four times that number.

After leaving the bridge, Cottam returned to the wireless room, put on the headphones, and waited for the set to warm up. The traffic had increased again, and he received some more ice warnings and messages bound for *Titanic* from Cape Race. He returned to the bridge and gave Dean the ice warnings, then returned, waiting once again for his set to warm up. He thought he would remind *Titanic* about the messages from Cape Race. He touched the telegraph key, the blue sparks cracking under his forefinger. Phillips came back, "Go ahead."

"Good morning old man (GM OM). Do you know there are messages for you at Cape Race?"[6]

There was a pause, punctuated by a flood of traffic, more than at any time Cottam could remember. Then Phillips blared through the traffic, his signal extraordinarily strong and loud. "CQD . . . CQD . . . SOS . . . SOS . . . CQD . . . MGY. Come at once. We have struck a berg. It's a CQD, old man (CQD OM). Position 41.46 N, 50.14 W."[7] Cottam felt a chill run down his spine. He couldn't move and stared at his notations. He felt the ship roll slightly, the whir of the transmitter, the tawny light of the incandescent bulb on the desk. He began to tap out a return message and asked if he should tell his captain. "Yes. Quick,"[8] came the fast reply.

Harold Cottam tore the headphones from his head and began a dead run for the bridge with the words forming in his mouth as he reached the bridge, telling First Officer Dean that the RMS *Titanic* has struck an iceberg and was sinking. Dean paused and then went down the ladder, passing though the chart room with Cottam right behind him, and burst into Captain Rostron's cabin. The captain looked up from his bunk clearly shocked that his first officer and wireless operator would barge into his cabin without even knocking. In his testimony later before Congress, Rostron related the scene. "The First Marconi Operator came to my cabin and came right up to me and woke me—well I was not asleep as a matter of fact and told me he had just received an urgent distress call from the *Titanic*—that she required immediate assistance—that she had struck ice."[9]

"Are you sure it is the *Titanic* that requires assistance?"

"Yes sir."

The captain raised himself up. The *Carpathia*'s range was limited to 130 miles on her wireless, and while it was understood that nighttime gave a wider range, there was still a great deal of room for error.

"You are absolutely certain?"

"Quite certain."

Rostron nodded and stood up.

"Alright. Tell him we're coming along as fast as we can."[10]

The captain stood up and looked at the first officer.

"Mr. Dean turn the ship around—steer northwest. I'll work out the course for you in a minute."[11] Dean left the room and Captain Rostron dressed quickly and headed for the bridge. Cottam ran for the wireless room. He sat down and pulled on the headphones, tapping out his response to the *Titanic*. "Putting about and heading for you."[12] The 13,564-ton liner had been meandering along the Atlantic passage from New York to the Mediterranean with close to eight hundred passengers divided between the three classes. Now all that was to change.

The forty-three-year-old captain left his cabin and went directly to the chart room. He worked out a new course then hurried to the bridge, giving the helmsmen the new course: North 52 West. He then called down to the engine room for full speed ahead. This was not enough for the "electric spark," as this would only give him 14 knots, which, by his calculations and the *Titanic*'s position, would put the *Carpathia* on a track to reach the stricken ship in four hours. Not good enough. Captain Rostron returned to the bridge and called up Chief Engineer Johnstone and told him he wanted every bit of speed he could coax out of the old liner.

He told the engineer to get every stoker out of bed and put them all on to fire up the furnaces under the boilers. Steam. This was the age of steam and steam was the rocket fuel of the day. Rostron needed every ounce of energy for his engines. He told Johnstone to shut off all the heat and hot water for the passengers and put all the steam into the reciprocating engines. Johnstone knew his duty and went below.

Then Captain Rostron gave First Officer Dean a list.

All routine work knocked off, the ship prepared for a rescue operation; swing out the ships boats; have clusters of electric lights rigged along the ships sides; all gangway doors to be opened; with block and tackle slung at each gangway; slings ready for hoisting injured aboard and canvas bags for lifting small children; ladders prepared for dropping at each gangway, along with cargo nets; forward derricks to be rigged and topped, with steam in the winches for bringing luggage and cargo aboard; oil bags readied in the lavatories to pour on rough seas if needed.[13]

Captain Rostron sent the first officer on his way and then called up the three surgeons aboard and gave them specific duties. Dr. Francis Edward McGee would handle the people in first class, with the Italian doctor overseeing second class and the Hungarian doctor tending to the third class. Stimulants and restoratives were to be set up at first aid stations in the dining rooms. Purser Brown was then instructed to make sure his stewards were stationed at the gangways as the *Titanic* crew and passengers boarded, to take down their names and classes and then to direct them to the proper first aid stations in the dining rooms.

Henry Hughes, the chief steward, was told to call out every crewman and that "soup, coffee, tea, brandy, and whiskey should be ready for those rescued."[14] Hughes was to oversee the conversion of the lounges, smoking room, and library into rooms for the survivors. He was then to put all of *Carpathia*'s third-class passengers into one area and free up space for *Titanic*'s steerage survivors. Rostron then made the decision that the *Carpathia* passengers were to be told nothing and were to be kept in their cabins. They didn't need people milling about while they conducted a rescue operation. To that end he had stewards stationed in the corridors to turn the ship's passengers back to their cabins. A master of arms, inspector, and several stewards would keep the third-class passengers in check. Panic was the last thing Captain Rostron wanted on his ship.

Down in the belly of the ship the stokers shoveled coal into the furnaces as the safety valves were closed and all the steam was redirected from the ship into the reciprocating engines pushing the pistons faster and faster. The *Carpathia*'s bow lifted, parting the freezing water with two long white streamers as her screws turned faster and faster, breaking past her registered speed of 14 knots and heading up to 17. The ship had never gone so fast, and the vibration of the engines shook the china in the dining rooms. Every passenger on *Carpathia* knew something was going on, as bedside glasses of water jumped and clothes gently swayed on racks. But the crew would tell them nothing as the curious few opened their doors. *Carpathia* was hauling through the darkness with men straining their eyes to see an iceberg that could easily send them down into the depths of the ocean.

Having done all that he could for the moment, Captain Rostron returned to the bridge and immediately posted extra lookouts. Taking the old ship beyond her limit in speed while passing though ice fields at night was dangerous, and the *Carpathia* could easily suffer the same disaster that had befallen the *Titanic*. He put an extra man in the crow's nest with two more lookouts in the bow and extra men on both wings of the bridge. He then put Second Officer Bisset, who was prized for his keen eyesight, on the starboard side of the bridge. No radar. No sonar. No lights to speak of. Captain Rostron was risking it all to go to the aid and rescue of a ship sinking in the North Atlantic in the middle of the night. They would need more than just keen eyesight and the efficiency of a good captain to make it to the *Titanic* without smashing into an iceberg themselves. Captain Rostron took a step toward the back of the bridge and lifted his cap off his head a few inches and closed his eyes, mouthing a silent prayer.

CHAPTER THIRTEEN

Signals

110 Minutes

IT WAS A RAINY, COLD NIGHT IN PITTSBURGH. WINTER HAD GIVEN WAY
to a soggy spring with overcast days and a persistent rain. The calendar
on the kitchen wall above the icebox and the wood-burning stove showed
the date as Sunday, April 14, 1912. It was cozy in the kitchen with the
checkered curtains. The low murmurs of the young men huddled around
the kitchen table with the bulky wireless set didn't reach to the two adults
upstairs. It was well after midnight and Monday was a workday. The over-
sized metal headphones on the young men looked otherworldly. Model
Ts still shared the street with horses, and most homes still had a barn in
the back. There was no radio, but there was wireless telegraphy, and that
was the purview of young men, the inventors, the men who had nothing
better to do than sit around with headphones on their ears listening for
distant beeps and buzzes.

Young men in America often had wireless telegraphy sets in their
kitchens and basements, stringing wires across roofs, across chimneys,
along gutters, in trees. Some people got creative and took pipes and made
what would later resemble television antennas. Some draped wires across
bushes. Anything to capture an elusive signal out of the ether coming in
from the coast or, even more amazingly, from the ocean.

Twenty-year-old Samuel Kerr sat in his kitchen in Pittsburgh along
with his friends "William Staving of Middle Street Northside and Frederick L. Nuttleman, 710 Middle Street Northside."[1] They were all amateur wireless operators, and they would gather in Kerr's kitchen around

his set to see what signals they could pull out of the night sky. On this night they had been listening in to some transmissions from New York and Boston. Then the weather changed as a front moved in. Light rain fell as the temperature dropped. Two hours later, the night was clear and cool, with pinpoint stars and a crescent moon over the Kerr home.

Kerr looked out the window and felt like they might get some far transmission tonight, maybe even a thousand miles. People didn't believe him that he had heard from Russian steamers, British ships, even distant stations in Newfoundland. No one understood the way the signals bounced off the ionosphere at night and winged their way into the interior of the United States. Kerr didn't mind if people didn't believe him because he knew what he had heard and he hoped one day to hear something astounding, a bit of news that only he would be the possessor of and would prove that he wasn't some crazy kid messing around in his mother's kitchen.

He had already proved the power of his wireless when he picked up the signals from the ship *Republic* when the *Florida* rammed the *Republic*, and Jack Binns, the wireless operator, had sent a distress signal. It was the first wireless success story, and Kerr had picked up the signal from Binns requesting assistance from more than a thousand miles away. It was simply unbelievable, but he knew before anyone else that the *Republic* had been rammed by the *Florida*. That was the farthest signal he had ever received. Now the static and the buzzes had been coming in in short bursts. Kerr adjusted his set. He had strung wire up between the chimney and one of the ventilation pipes in the roof. His father had shaken his head when he saw the wire. Kerr had gone up to the roof several times to adjust the wire and tried different configurations to see if he could improve his reception. Tonight, however, the amount of traffic let him know that his latest adjustments were bearing fruit.

He waited with his friends clustered around the kitchen table. The house was quiet, and Kerr was about to adjust his set again when the static cleared. He jotted down three dots, three long dashes, then three dots. He felt his heart jump. It was an SOS. The universal distress signal was brand new and was supposed to replace CQD. He wrote SOS on his pad and William's and Fred's eyes lit up. The signal repeated itself, then bursts of static moved in and cleared once again before the Morse came flooding in. Kerr wrote the letters down quickly. *SOS CQD MGY Here corrected position 41.46 N, 50.14 W. Require immediate assistance. We have collision with iceberg. Sinking.*[2]

Kerr scribbled down the message relayed from Cape Race and stared at his friends. MGY MGY. He knew the calls letters. He had read about the ship with the most powerful wireless ever installed. It was the largest ship in the world and had just left on her maiden voyage and it was all over the press. As a wireless operator, Kerr had found out the equivalent of the ship's phone number, MGY. It was the *Titanic*. He stared at his friends and flipped the pad around with his headphones still on, mouthing the words in the small kitchen in Pittsburgh. *SOS CQD MGY Here corrected position 41.46 N, 50.14 W. Require immediate assistance. We have collision with iceberg. Sinking.*[3]

Kerr pulled off the headphones and stared at the pad and then his friends. The kitchen clock ticked in the silence. Signals from out in the North Atlantic had just entered the warm kitchen in a suburb of Pittsburgh in the year 1912. For a moment, the three friends were there in the freezing cold along with the thousands of people facing the prospect of drowning in 25-degree water. Samuel Kerr felt a chill go down his spine. A Marconi operator on the RMS *Titanic* had sent out a frantic SOS saying the RMS *Titanic* had struck an iceberg and was sinking. Kerr put back on the headphones and the signals kept coming. *SOS CQD MGY Here corrected position 41.46 N, 50.14 W. Require immediate assistance. We have collision with iceberg. Sinking.*[4] He jumped up and ran to tell his sleeping parents upstairs that the largest ship in the world had hit an iceberg and was about to sink.

Chapter Fourteen

Sinking, Can Hear Nothing for the Steam

105 Minutes

Jack Phillips braced himself against his table. The slant of the *Titanic* had already cleared his desk of pens and coins, and his pad had slid off twice. The light over his desk was now hanging at an angle and had become a crude clinometer telling him the angle of the ship's slant toward the sea. It was like working on the edge of a cliff, and the key had slipped out from under his hand several times. The power was still on and the dynamos were still running, but the lights had a strange, dim, reddish glow now.

Phillips had been transmitting nonstop for the last hour, and in a giant circle the signals bounced from the *Titanic* and went to the *Parisian* (45 miles away but her wireless operator asleep), the *Mount Temple* (less than 50 miles away), the *Californian* (10 miles away, wireless operator asleep), the *Birma* (70 miles away), the *Carpathia* (50 miles away), the *Baltic* (200 miles away), the *Frankfurt* (140 miles away), and the *Virginian* (170 miles away), then swinging around simultaneously to shore stations at Sable Island, Cape Race, Camperdown, and Cape Sable. These signals were not received as in a telephone but were passed on, relayed, and listened to simultaneously, and if one drew lines between the ships, the *Titanic*, and the shore station, it would be a giant trapezoidal party line. Other ships that were too far to offer assistance picked up the signals as well, the SS *Prinz Friedrich Wilhelm*, the RMS *Caronia*, the RMS *Celtic*, the SS *Cincinnati*, and *Titanic's* sister ship, the *Olympic* (400 miles away).

The problem was that the position given by Captain Smith to Jack Phillips of 41.44 N, 50.24 W was wrong. This was corrected to 41.46 N, 50.14 W, but it was still off. In 1912, a sextant was used to determine latitude by sighting on the North Star and measuring the angle between the star and the northern horizon. Longitude was harder though. It relied on identifying a star touching the eastern or western horizon and then measuring the angle between the horizon and the imaginary line drawn from the star to the North Star. *Titanic* was actually at 41.44 N, 49.56 W, which was thirteen miles east and two miles south of where Captain Smith or First Officer Murdoch had determined. It was only in 1985, when the wreck was discovered, that this error in navigation was discovered.

Still, if any ship could get close to the *Titanic* even with the error, then they would spot her rockets or see the lifeboats or be guided closer by Phillips on the wireless. The night was uniquely clear, and unobstructed light traveled for miles. Rockets on such a night high in the coal-black sky could easily be seen for thirteen miles. But there was another problem as well. When Phillips or Harold Bride held down the key to their spark-gap transmitters it used a modulation method called damped waves. As long as the key was pressed down, there would be a series of pulses or radio waves that repeated at audio rate at about 50 to several thousand Hertz. In the headphones of the receiving operators, this sounded like a buzz. The duration of it separated dots from dashes. But damped waves occupied a wide band of frequencies and interfered with other frequencies. Consequently, it became a war of who had the stronger transmitter, with ships like the *Titanic* drowning out less powerful wireless sets. But a ship could be close by and also wreak havoc on the process, stepping all over the transmissions of another. This meant there were errors from interference as when Phillips told the *Californian* to shut up and keep out when he was trying to transmit passenger messages to Cape Race, and a similar situation happened with the *Frankfurt*. The damped wave transmissions stepped on each other and messages were lost in transmission.

Phillips had been hunched over tapping out CQDs and SOSs now for almost an hour, and the *Carpathia* had just told him they were coming fast. She was only fifty miles away and might just reach them in time, although the angle of the *Titanic* told him otherwise. He began sending out a new distress call with the new position Captain Smith had given him. "MGY TITANIC CQD Here corrected position 41.46 N, 50.14 W. Require immediate assistance. We have collision with iceberg. Sinking. Can hear nothing for noise of steam."[1] He repeated the message twenty times.

Operator John Durant on the *Mount Temple* had heard the conversation between the *Carpathia* and the *Titanic* and once again tried to establish contact, but the strength of his signal was too weak to break in. The *Prinz Friedrich Wilhelm* heard Phillips's last signal, which gave his position as 39.47 N. 50.10 W. "Are you coming to our assistance?"[2] asked Phillips. "What is the matter with you?"[3] came back. Phillips told the German operator, "We have collision with iceberg. Sinking. Please tell captain to come."[4] The German wireless operator promised to do so, then went to find his captain, and even though the ship was too far to make a difference she turned around and headed for the *Titanic*. Phillips then sent out a different signal. He was trying to capture anyone who might lend assistance and the international call had to be translated, but he didn't know what might communicate their position the most effectively. "I require assistance immediately. Struck by iceberg in 41.46 N, 50.14 W."[5]

The RMS *Caronia* picked up the faint message and relayed to ships in the area, repeating Phillips's call for help. The *Mount Temple* tried again when Phillips paused in his transmission, and the *Mount Temple* finally broke in with, "Our Captain reverses ship. We are about 50 miles off."[6] It was now 12:30 a.m., and things were looking up.

The wireless operator William James Cotter on the *Virginian* was 170 miles away and picked up the *Titanic*'s distress call as well. He went running for the bridge where he handed the first officer the message. The *Titanic* was famous and rumored to be unsinkable. The officer glared at Cotter. It was a joke, he decided, and ordered the young operator off the bridge. Cotter refused to move and stated that the message was true. Two deck hands lifted him off the deck and carried him toward the stairs leading back to the radio room. Cotter broke away and ran for the captain's door and began kicking the door and slamming it with his fist. When Captain Gambell opened the door, groggy with sleep, Cotter breathlessly told him he had received the message that the *Titanic* had struck an iceberg and was sinking. Captain Gambell stared at the heaving wireless operator and the first officer standing behind him. He ordered the ship turned around immediately and to head for the *Titanic*'s position at full speed.

Phillips kept transmitting with Bride assisting him and running messages to Captain Smith. The news of the *Carpathia* steaming full speed had been one of the messages Bride placed in Smith's hands. *CQD SOS CQD. MGY Sinking . . . request immediate assistance.* The world flickered and dimmed in the wireless cabin with the overhead lightbulb swinging closer and closer to the wall as *Titanic* rose higher. Phillips's hand ached. *CQD SOS CQD MGY Sinking . . .*

Chapter Fifteen

The Light

100 Minutes

Lawrence Beesley was hovering seventy feet up in the darkness in his lifeboat. The boat made its way jerkily down to the sea. So far, for Beesley, the evacuation had gone smoothly, with the lowering of the lifeboat being one more drama in the staid life of an English schoolteacher.

> *It was exciting to feel the boat sink by jerks, foot by foot, as the ropes were paid out from above and shrieked as they passed through the pulley blocks, the new ropes and gear creaking under the strain of a boat laden with people and the crew calling to the sailors above as the boat tilted slightly, now at one end, now at the other, "Lower aft!" "Lower stern!" "Lower together!"*[1]

Beesley was in a lifeboat loaded with sixty people, one of the few boats that had been loaded to capacity. Down, down, down, the boat slowly descended into the frigid night. He was still enjoying the view. "It certainly was thrilling to see the black hull of the ship on one side and the sea, seventy feet below . . . or to pass down by cabins and saloons brilliantly lighted."[2] But the thrilling drama Beesley was observing quickly turned dire when the boat approached a gushing funnel of water from the side of the *Titanic*. It was the condenser exhaust just above the waterline; a fire hydrant of warmed water blasting out of the side of the ship, and Beesley's boat was headed right for it. A crewman shouted. "We are just over the condenser exhaust: We don't want to stay in that long or we shall be swamped. . . . Feel down on the floor and be ready to pull the pin which lets the ropes free."[3] Essentially, Beesley and the other passengers were coming down into the stream of the gushing water. Beesley had noticed the water before. "So large was the volume of water that as we ploughed along and met the waves coming toward us, this stream would cause a splash that sent spray flying."[4]

So now Beesley's boat is in the water but still attached to the *Titanic*. The pin was never found, and the boat had landed near the expelling water with the force driving the boat away and the ropes pulling it back against the ship. This boat was stuck and in real danger of sinking. Beesley then looked up and saw another boat, Boat 15, coming down right on top of them. "Looking up we saw her already coming down rapidly from B Deck; she must have filled immediately after ours."[5] Beesley and others shouted up to stop lowering, but still the boat came down. "She dropped down foot by foot—twenty feet—fifteen—ten—and a stoker and I in the bows reached up and touched her bottom swinging above our heads, trying to push away our boat from under her."[6] Not only was Beesley's lifeboat now in danger of being swamped from the excess water of the boilers but was also about to be crushed from another lifeboat coming down directly on top of them. "It seemed now as if nothing could prevent her dropping on us, but at this moment another stoker sprang with his knife to the ropes that still held us, and I heard him shout, 'One! Two! As he cut them through!'"[7] The boat swung away as Boat 15 dropped in place, just missing the passengers of Boat 13. The force of the condenser stream then washed Beesley's boat clear of *Titanic*, the oars were manned, and they made a fast getaway from what seemed certain disaster.

The lack of organization by the *Titanic* crew and officers in loading the lifeboats was immediately plain to the schoolteacher:

The crew was made up of cooks and stewards, mostly the former . . . their white jackets showing up in the darkness as they pulled away, two to an oar. I do not think they had any practice in rowing for all night long their oars crossed and clashed . . . shouting began from one

of the boats to the other as to what we should do, where we should go,
and no one seemed to have any knowledge how to act.[8]

The stoker who had cut the boat free was elected as the captain of Beesley's boat, but they still had no direction. "Not that there was anywhere to go or anything to do."[9]

Quickly, however, Beesley and others saw the lights on the horizon that Second Officer Lightoller and Jack Thayer's mother had seen. "We saw what we all said was a ship's lights down on the horizon on the *Titanic's* port side: two lights, one above the other, and plainly not one of our boats; we even rowed in that direction for some time, but the lights drew away and disappeared below the horizon."[10] They stopped rowing, and Beesley looked up into the night sky of brilliant stars reflected on the glass of the ocean. The schoolteacher's literary skills are on display here, but it does give a picture of what it was like to be in the middle of the dark Atlantic on the frozen night of April 14, 1912.

> *The night was one of the most beautiful I have ever seen: without*
> *a single cloud to mar the perfect brilliance of the stars, clustered so*
> *thickly together that in places there seemed almost more dazzling*
> *points of light set in the black sky than background of sky itself. . . .*
> *They seemed so near and their light so much more intense than ever*
> *before, that fancy suggested they saw this beautiful ship in dire distress*
> *below and all their energies had awakened to flash messages across the*
> *black dome of the sky to each other; telling and warning of the calam-*
> *ity happening in the world beneath.*[11]

Quartermaster Robert Hutchins in Boat 6 was also rowing toward the light on the horizon. Lightoller had directed him toward the light, and he kept rowing, watching the *Titanic* as he rowed. The *Titanic* was in a death tilt now, with the port lights beginning to touch the water. Hutchins could not get over how still the night was and how the water mirrored the stars. The light that Lightoller had directed him to had been the reason the rockets were bursting overhead now. But as hard as he rowed, it seemed the light was always farther away. Lightoller himself had been watching the light, trying to get the ship's attention. He had been Morsing the ship, shooting off rockets, watching her to see if anything came back. Nothing did. He watched the ship with his binoculars, not realizing there was a man watching him. On the *Californian*, Second Officer Stone relieved Third Officer Groves at midnight on the bridge. Stone ran into Captain Lord in the wheelhouse as he was making his way to the bridge. Lord pointed out the steamer to Stone, and the second officer "noticed that she was displaying her red port sidelight, along with a single masthead light and a few smaller lights; she also seemed to be showing a lot of deck lights."[12] Stone thought the glow from her lights was so strong that he estimated her to be no more than five or six miles distant. Lord told Stone to notify him if anything changed or she drew closer to the *Californian*. He then headed for the chart room.

Groves went down to the wireless office to chat with the operator, Evans, who was lying in his bunk with a magazine. Groves liked to pick Evans's brain about the intricacies of wireless and felt he was learning Morse code at a good pace. Sometimes he listened in to catch some news, but Evans seemed quiet and not in the mood for conversation. Groves picked up the headphones and slipped them on. He heard nothing.

"What ships have you got, Sparks?"

"Only the *Titanic*," he replied.[13]

Groves listened for a while longer not understanding the set had a clockwork-driven magnetic detector and had to be wound up. So, he heard nothing and laid the headphones back down and left. Up on the bridge Stone had been watching the glittering ship on the horizon. Groves had said he had tried to contact the stranger with the Morse lamp but was unsuccessful. Apprentice Officer James Gibson came up on the bridge and bought Stone some coffee. Stone pointed out the ship to Gibson and remarked they had been unable to make contact. Gibson lifted his binoculars and stared at the ship. He "could clearly make out her masthead light, her red sidelight, and the glare of white lights on her after decks."[14] Gibson concurred it was a large ship and then left the bridge. Stone stared into the night walking the bridge slowly. At 12:40 a.m. the captain called up the voice tube and wanted to know if the ship had come any closer. Stone replied it had not.

He continued walking the bridge, keeping an eye on the liner.

Around 1 a.m. . . . Stone was startled by a flash of white light above the other ship. Taken by surprise and unsure of exactly what he had seen, he watched the stranger closely and, after a few minutes, was awarded with another white flash—and this time he was able to identify what he was seeing: a white rocket bursting high above the unknown vessel, sending out a shower of white stars.[15]

Second Officer Stone counted five rockets and watched as more exploded over the ship. He kept watching until 1:30 a.m. and decided to let Captain Lord know through the voice tube. Lord was napping on the settee in the chart room and groggily answered. Stone told him what he had seen.

"Are they company signals?"

"I don't know sir."

"Well go on Morsing."

"Yes sir."

"And when you get an answer let me know by Gibson."

"Yes sir."[16]

Gibson returned to the bridge and Stone told him about the rockets. Gibson trained his binoculars on the ship and saw a fiery white explosion light the sky. They watched the rockets exploding over the ship. Stone borrowed Gibson's glasses and then handed them back. "Have a look at her now. She looks very queer out of the water—her lights look queer."[17] Gibson stared through the lenses and the ship appeared to be listing. "A big side out of the water"[18] was the way he would later describe it. "A ship is not going to fire rockets at sea for nothing,"[19] Stone murmured. Gibson nodded slowly staring at the glittering ship.

"There must be something wrong with her."[20]

There was a lot wrong with the *Titanic*, and Lightoller could not understand why the steamer on the horizon was not reacting to the rockets or the Morse. The boats were dropping more quickly now. Captain Smith himself had told some of the boats to row for the light.

Lightoller told some reluctant women, "Do you see her—that ship over there. Probably she's a sailing ship. The breeze should spring up at daybreak, and she'll be able to come and pick us up from the boats."[21] Still, even with the rockets and the listing ship, the officers aboard the *Titanic* were having a hard time understanding that the ship was really sinking. Fourth Officer Joseph Boxhall on the bridge stood next to Captain Smith and leaned over. "Is it really serious?"[22] Captain Smith didn't move but answered quietly. "Mr. Andrews tells me that he gives her from an hour to an hour and a half."[23]

No one had to tell Charles Lightoller the ship was sinking. He had been watching the water creep up the emergency stairs that ran from E Deck to the boat deck. The glowing green water was covering the stairs methodically and this was Lightoller's gauge on how much time they had left. The first-class passengers boarding the lifeboats had been calm, but the extreme tilt of the ship, the exploding rockets, and the departing lifeboats now produced a creeping insidious panic on the ship.

Jack Thayer and his friend Long had been separated from his father and mother and were roaming the deck when he noticed a scuffle in the last two forward boats and saw the director of the line, Bruce Ismay. Ismay had been standing all night by the boats urging people and crew. His more subdued tone made him blend in as a handsome, mustached gentleman in a dark suit coat with the edges of his pajamas sticking out from under his pants and over his high laced shoes. Ismay had been Captain Smith's right-hand man during a lot of the voyage, sitting at his table in the smoking room in his dinner jacket, taking ice warnings, setting the arrival time to New York for Wednesday morning instead of Thursday night. Now he was one of the many contemplating death in the freezing Atlantic. Boat 14 was being loaded, and like a silent acrobat he deftly climbed in and sat down. No one told him to get into the boat. He simply did it and joined the forty-two people. Bruce Ismay wanted to live.

But he was not unobserved. Lillian Minahan was sitting in Boat 14 and later wrote, "Then something that made my blood cold happened. The lifeboat was held a moment—J. Bruce Ismay was clambering into it, and he was being assisted by a couple members of the crew. There was a dreadful expression on his face as he took a place in the already over-crowded boat." Thayer later wrote about the final boats:

> A large crowd of men was pressing to get into them. No women were around as far as I could see. I saw Ismay, who had been assisting in the loading of the last boat, push his way into it. It was like every man for himself. . . . Purser H. W. McElroy as brave and as fine a man as ever lived, was standing up in the next to last boat, loading it. Two men, I think they were dining room stewards, dropped into the boat from the deck above. As they jumped, he fired a revolver twice into the air. I do not believe they were hit, but they were quickly thrown out.[24]

The *Titanic* mythology is one of order. Shots may have been fired, but throughout the years the story has held that the only shots fired were into the air. Yet personal letters from survivors tell a different story. When the third-class passengers burst onto the upper decks and found that most of the boats had already left, one can imagine the panic and the fight for survival to get into any boat that remained. A letter from third-class passenger Eugene Daly to his sister in Ireland recounts a tragic incident. Written soon after the sinking, on May 4, 1912, the letter appeared in the *Daily Telegraph*:

> At the first cabin when a boat was being lowered, an officer pointed a revolver and said if any man tried to get in, he would shoot him on the spot. I saw the officer shoot two men dead because they tried to get in the boat. Afterwards there was another shot, and I saw the officer himself lying on the deck. They told me he shot himself, but I did not see him. I was up to my knees in the water at the time. Every-one was rushing around and there were no more boats. I then dived overboard.[25]

This makes more sense than the brandy and cigars scenes history has handed down of gentlemen bravely facing their fate. The primal urge to survive overrides all else. The prospect of thousands of people facing drowning in the icy Atlantic is at odds with the band playing merrily while men in coat and tails bid lovers, wives, daughters, farewell.

Yet, the breakdown of order is not something that has come down to us in the histories of the *Titanic*. It has been the direct opposite, with the gallantry, chivalry, and moxie of mostly white Anglo-Saxon Protestant men dominating the narrative. Eugene Daly did survive, and one letter does not confirm the event, but in 1981 another letter surfaced from first-class passenger George Rhemis, written to his wife on April 19, 1912. Rheims jumped overboard, but what he wrote does seem to back up Daly's account of mayhem aboard the *Titanic*:

> *While the last boat was leaving, I saw an officer with a revolver fire a shot and kill a man who was trying to climb into it. As there remained nothing more for him to do, the officer told us, "Gentleman each man for himself. Goodbye." He gave a military salute and then fired a bullet into his head. That's what I call a man!*[26]

It is hard to believe two different men writing private letters to loved ones would fabricate such a story. These have the ring of truth and break up the patina of the *Titanic* myth that would have the patriarchal Edwardians setting a tone of grace and order in the face of certain death.

It gets worse. Dr. Washington Dodge, a first-class, fifty-one-year-old passenger who would suffer a nervous breakdown seven years after the *Titanic* sank and eventually commit suicide, told the *San Francisco Bulletin* on April 19, 1912,

> *Some of the passengers fought with such desperation to get into the lifeboats that the officers shot them and their bodies fell into the ocean. . . . As the excitement began, I saw an officer of the* Titanic *shoot down two steerage passengers who were endeavoring to rush the lifeboats. I have learned since that twelve of the steerage passengers were shot altogether, one officer shooting down six.*[27]

Dr. Dodge also saw First Officer William Murdoch shoot himself. He observed this from his lifeboat where Murdoch was by the forward lifeboat station on the starboard side.

> *We could see him from the distance that two boats were being made ready to be lowered. The panic was in steerage, and it was that portion of the ship that the shooting was made necessary. Two men who attempted to rush beyond the restrain line were shot down by an officer who then turned the revolver on himself.*[28]

Of course order was breaking down. *People were fighting for their survival.* It is amazing that the glossy brush of the Edwardian era has managed to suppress these stories in the mainstream of *Titanic* archeology. Twelve third-class passengers shot down while Bruce Ismay slips into one of the remaining boats. This does not surface in most of the histories of the *Titanic.* This was anarchy, primal, and it had none of the Gilded Age patina that has been ascribed to the code of conduct on the *Titanic.* The third-class passengers were being cut down as they fought for survival. Pandemonium was a creeping presence as the last few boats were lowered away. A young French girl stumbled climbing into Boat 9 and missed the boat entirely, falling between the bow and the ship. Someone grabbed her by her ankle and she dangled sixty feet in the air before passengers pulled her into the promenade deck. Another woman became hysterical as she climbed into Boat 11 as Steward Witter stood on the rail assisting her when they both fell into the boat. A large woman shrieked as she neared Boat 13, "Don't put me in that boat. I don't want to go in that boat. I have never been in an open boat in all my life!"[29] Steward Ray grabbed her by the arms. "You've got to go, and you may as well keep quiet."[30]

There was a shortage of seamen as the boats pulled away, and when Boat 6 was halfway down a woman called out that they needed another seaman. "We've only one seaman in the boat." Lightoller turned around. "Any seamen there?" Major Peuchen stepped forward. "I'm not a seaman, but I'm a yachtsman, if I can be any use to you." By this time, the boat was nine or ten feet down toward the sea. Lightoller stared at the man.

"If you're seaman enough to get out on those falls and get down into the boat you may go ahead."[31] The vice commodore of the Royal Canadian Yacht Club swung himself out on the forward fall and slid down into the boat. He was one of the few males Lightoller allowed in a boat that night. On the starboard side, the rule of women and children seemed to have more flexibility. Officer Murdoch was allowing first-class male passengers into boats if there was room.

French Aviator Pierre Marechal and sculptor Paul Chevre climbed into Boat 7. A couple of Gimbles buyers reached Boat 5. When the time came to lower Boat 3, Henry Sleeper Harper not only joined his wife, but he brought along his Pekinese Sun Yatsen and an Egyptian dragoman named Hamad Hassah, whom he had picked up in Cairo as sort of a joke.[32]

But the first class had a big advantage over the second- and third-class passengers. They greased the palms of the stewards, and many had stewards who had been loyal to them and served families on other ships through the years. Dr. Dodge was loitering around Boat 13 when Steward Ray saw him and asked if his wife and son had boarded a boat. Ray had served the Dodges on the *Olympic* and had even suggested they take passage on the *Titanic's* maiden voyage. He was relieved the doctor's family was safely in boats, since he was partly responsible that they were even on the ship. He went and stood by Boat 13. "You had better get in here," he called out, pushing the doctor into the open space."[33] The most famous scene of sycophancy occurred with Sir Cosmo Duff-Gordon, along with his wife and her secretary. Duff-Gordon was near Boat 1 and asked Murdoch if he might enter. "Oh, certainly do. I'll be very pleased,"[34] the *Titanic* officer said, practically bowing. He then let in some Americans and C. E. H. Stengel, a portly man who fell over getting into the boat and gave Murdoch a chuckle, murmuring, "That's the funniest thing I've seen tonight."[35] He then added six stokers and lookout Symons, instructing him, "Stand off from the ship and return when we call you."[36] This was for the passengers, as Murdoch knew there was no returning at this point.

Boat 1 became the poster child for all that was wrong with the loading of the lifeboats. There were twelve people for a boat that held forty. This was repeated in varying degrees in every boat, and hundreds of people lost their lives because the boats were never loaded to their rated capacity. But it would get worse as the night unfolded. Jack Thayer and his friend had still not found a boat.

Long and I debated whether or not we should fight our way into one of the last two boats. We could almost see the ship slowly going down by the head. There was so much confusion, we did not think they would reach the water right side up and decided not to attempt it. I do not know what I thought would happen, but we had not given up hope.[37]

Thayer and Long watched the last boat being lowered and realized no one was directing the bow and the stern to keep the boat level. "The boat was lowered so fast that the people were almost dumped out into the water. I think if Long and I and others had not yelled up, 'Hold the bow,' they would have all been spilled out."[38] The third-class passengers were now flooding up from below and merging with the people running from the creeping water. The *Titanic* was slowly rising up like a gigantic teeter-totter with the water malevolently creeping toward the stern, where Jack Phillips and Harold Bride were frantically calling for help.

Chapter Sixteen

Olympic

95 Minutes

TITANIC'S TWIN, THE OLYMPIC, CHURNED THROUGH THE COLD ATLANTIC. Like most people traveling on the high seas at night, her passengers were asleep or finishing up one last cigar and one last brandy in the smoking rooms. One man who was still up was Ernie Moore, tapping his telegraphic key in the wireless room. The room was on the highest deck of the ship, the boat deck, and forty feet behind the bridge. It was a no-frills wireless room, lacking even a porthole to offer a view onto the crystal-clear night, but there was a skylight that poured sunlight down during the day. The *Olympic's* wireless operators seldom left the room— they were simply too busy. The ship offered twenty-four-hour wireless service, and Moore and junior operator Alec Bagot were constantly at their post transmitting passenger messages, the business of White Star, and the ice warnings and navigation reports that went to the bridge. Both the *Titanic* and *Olympic* had been outfitted with the most powerful Marconi wireless telegraphs available, giving the *Olympic* a range of 250 miles during the day and up to 2,000 miles at night.

The *Olympic*, like the *Titanic*, was outfitted with a squash racquet court, gymnasium, a pool with cabins decorated in a variety of styles, fireplaces, electric lights, and heaters. The promenade deck ran the ship's length, and the band played at every meal in the formal dining room, which extended the full ninety-two-foot width of the ship. Still, to travel on the *Olympic* was but a prelude for the final symphony of luxurious transatlantic passage on the *Titanic*.

It was now Sunday, and the shipboard activities had wound up early. Most of the passengers had retired to their cabins. Moore's shift in the wireless room was about to end and he was getting the news from Cape Race for the morning newspaper published onboard. Passengers would wake to find the latest news, sports, and weather, courtesy of the two young men who worked throughout the night. There was nothing much different from the news they had received the day before in New York. It was just a practice among all the ships to tune into the shore stations like Cape Race and get news for their shipboard publications. At 12:40 a.m., Moore turned back to the commercial wavelength for ship-to-ship messages.

A ship with the call sign MGY was calling to all ships, and Moore instantly recognized it as the *Titanic*. It was signaling about some ship that had struck an iceberg. The static overwhelmed the signal and too many other ships were signaling at the same time. The senior wireless operator wasn't sure if the *Titanic* was signaling for herself or someone else. Moore transmitted over the chatter and asked if *Titanic* needed help. Then he listened for a full ten minutes before the MGY call letters came again. "I require immediate assistance. Position 41.46 N, 50.14 W."[1] Moore paused, then shot back, "Message received." He didn't bother waking the junior operator, Bagot, but wrote down the message and put it in a sealed envelope and headed for the bridge.

Moore handed the message to the officer of the watch and requested it be given to the captain immediately. The fifty-one-year-old Captain Herbert J. Haddock believed what most did about the White Star super-liners, that the *Olympic* and the *Titanic* were immune to the dangers of the sea, and he was in shock when he read the message that the *Titanic* needed immediate assistance. Captain Haddock had just taken over command of the *Olympic* on April 3rd in Southampton and taken her to New York. This was his first trip back to Europe from America on the *Olympic* and he was due to pass the *Titanic* in mid-ocean. The connection between the two captains involved more than their current voyages. Captain Smith had formerly commanded the *Olympic* and both men had taken over their new commands at the same time. Haddock had actually trained on the *Titanic* while she awaited sea trials, and although he had never taken her out, he came to understand the ship.

Now the captain of the *Olympic* was in the chart room where Captain Smith had stood a few weeks before, and he was working out a course to reach the *Titanic* as quickly as possible. His own position was 40.52 N, 61.8 W. He determined he was a good five hundred miles from the *Titanic*. Ten minutes after he received the wireless message, he had the ship turn to the new coordinates and ordered the engineer put every bit of steam into *Olympic's* massive reciprocating engines. All stokers were woken up and put to work, driving the ship to a higher speed than she ever had achieved before, jumping from 19 knots to 23. "Men worked as they never had before, even to the verge of breaking down."[2] Captain Haddock was late in getting the information. The *Titanic* had sent out her first CQD at 12:15 but the *Olympic* did not pick up the signal for another twenty-five minutes. One hour had already elapsed since the *Titanic* had struck the iceberg.

Operator Moore returned to his wireless cabin and found the chatter shooting all over the North Atlantic over the *Titanic*. It was highly annoying. Ships and wireless shore stations were stepping all over each other trying to get information. Moore signaled CQ QRT which meant all stations stop signaling. He then sent another message to the *Titanic*. "This is *Olympic* calling *Titanic*, stop talking stop talking."[3] Phillips replied at 1:10 a.m. and told the whole story. "We are in collision with berg. Sinking head down. 41.46 N, 50.14 W. Come soon as possible."[4] Then a second message. "Captain says, get your boats ready. What is your position?"[5] At 1:25, Moore sent his position and asked, "Are you steering southerly to meet us?"[6] Phillips came back immediately: "We are putting the women off in the boats." Then moments later: "We are putting the women off in small boats."[7]

Moore understood then how dire the situation was from this one sentence. Putting off passengers into the Atlantic Ocean meant she was sinking and sinking fast. He stood and opened the door to the sleeping quarters and shook junior operator Bagot awake. Moore went back to transmitting, each crash of the high voltage across the rotary gap cracking across the room. Bagot pulled his uniform over his pajamas and went and stood behind Moore, who handed him an envelope with "Commander *Olympic* Urgent the Bridge" on the outside. Bagot left for the bridge to find the captain with the message telling him that the *Titanic* was putting off passengers into small boats in the middle of the North Atlantic in the middle of the night. Bagot found Captain Haddock, First Officer Hume, Second Officer A. H. Fry, and two other junior officers on the bridge. He handed the envelope to the officer of the watch, who gave it to Captain Haddock. Haddock scribbled on a note pad: "Commander *Titanic* am lighting up all possible boilers as fast as I can—Haddock."[8]

Bagot rushed for the wireless room. The passengers aboard the *Olympic* were not to be informed. This was Captain Haddock's order. But American architect Daniel Burnham, who was famous as the driving force and designer behind the Columbian Exposition in Chicago of 1893, had already started to figure out something was not quite right. He had spent much of the voyage in his stateroom as he suffered from diabetes which gave him intense pain in his feet. He was on vacation and his friend and collaborator Frank Millet was on the *Titanic* traveling with Major Archibald Butt, the military aide to President Taft. It was logical that he would send him a telegram from the ship, and so he did.

Chapter Seventeen

Third Class

90 Minutes

Schoolteacher Lawrence Beesley, in Lifeboat 13, stared at the *Titanic* from the darkness of the ocean and was amazed that "she was absolutely still."[1] He later wrote,

> *Indeed from the first it seemed as if the blow from the iceberg had taken all the courage out of her, and she had quietly come to rest and was settling down without an effort to save herself, without a murmur of protest against such a foul blow. . . . She sank lower and lower in the sea, like a stricken animal.*[2]

Beesley describes the view from his lifeboat:

> *Imagine a ship nearly a sixth of a mile long, seventy-five feet high to the top decks, with four enormous funnels above the decks and masts again high above the funnels; with her hundreds of portholes, all her saloons and other rooms brilliant with light, and all around her, little boats filled with those who until a few hours before had trod her decks and read in her libraries and listened to the music of her band in happy content; and who were now looking up in amazement at the enormous mass above them and rowing away from her because she was sinking.*[3]

What Beesley could not see was the panic on the ship where the empty davits with their ropes and pulleys dangling spoke of the last lifeboats that had just left. The third-class passengers had flooded up onto the upper decks and, unlike the first-class passengers, they had no idea where they were even going. To begin with, third class, or "steerage," was in the bowels of the ship, so they had to make their way up to the boat deck or any deck just to have a chance to get into a boat. Men and women swarmed around the main steerage staircase which was aft of E Deck. The stewards had woken up the third-class passengers, yelling down the hallways, but then didn't tell them where to go. Initially it was women and married couples, but now the men arrived from the forward section via the long passage "Scotland Road," and the crowd grew with people jammed up against the white walls and under the bare lightbulbs.

Still, some stewards tried to help by getting them into life jackets. Steward John Edward Hart in third class confronted the basic problem when he tried to explain that the life jackets would save their lives: a large portion of the third-class passengers spoke no English. There were interpreters, but it was hard to make sense among the confusion and noise of frightened immigrants. At 12:30, the order for women and children to come up had filtered down to frightened people, and now began the odyssey that for many would lead back to where they had started.

The third-class section was traditionally sealed off from most of the ship. The class system of Britain was rigid. The *Titanic* was a maze; even experienced hands like Second Officer Lightoller would find himself turned around. He said it took him weeks to find his way around the ship. Here were people who didn't even speak English and had been restricted to the lower areas of the ship. Steward Hart knew they would never find their way, so he started off with little groups. Just organizing the groups took time. Add to this the fact that the ship was slanting downward, making the people more frightened, loud, and hard to communicate with. Finally, Hart's first group was ready.

> It was a long trip—up the broad stairs to the *Third-Class Lounge* on *C Deck* . . . across the open well deck . . . by the *Second-Class library* and into First-Class quarters. Then down the long corridor by the surgeon's office, the private saloon for the maids and valets of First-Class passengers, finally up the grand stairway of the boat deck.[4]

The steward made it to Boat 8 with his first group, but then another problem arose. When Hart put them into the boat, they ran back inside where it was warm. He made his way back to E Deck for another party, but women now refused to go, and men demanded to go. It is amazing that Hart was able to get another party up to the boat deck and into Boat 15. It was 1:20 by now. First Officer Murdoch ordered Hart into the boat, and he found himself at sea by 1:30. This was one of the few good stories of the third-class evacuation from the *Titanic*. There was no policy on how to get the third-class passengers up from the bottom of the ship. Many of the passengers struck out on their own and, avoiding the cul de sac on E Deck, somehow ended up topside, but this was not the boat deck. Many wandered off into other parts of the ship with no one to guide them. Many of the barriers separating the third class were still up and people found themselves blocked, having to go back down long corridors with no idea where they were headed. "Like a stream of ants, a thin line of them curled their way up a crane in the after well deck, crawled along the boom to the First-Class quarters, then over the railing and on up to the Boat Deck."[5]

Many were herded along and reached the second-class promenade deck, but then found they could go no farther. An emergency ladder for crew use became their escape and for many it was the first time they were able to gaze into a first-class restaurant through the windows near the ladder. The tables gleamed with silverware, fine china, and candelabra. For most, the opulence was stunning. Class erosion would begin here as the restraints were falling apart and, like the *Titanic*, the old world began to collapse. The carefully sectored Edwardian worlds shattered as the third-class passengers, now frantic to get off the ship, smashed the barriers meant to keep them segregated from the extremely fortunate. "As third-class passenger Daniel Buckley climbed some stairs leading to a gate to first class, the man ahead of him was struck down by a seaman standing guard."[6]

The passenger then got up and charged up the stairs again as the seamen locked the gate and left. The third-class passenger broke the lock and the gate fell, with Buckley and the rest of the third-class passengers swarming into first class like the ocean that also was making its way up from below. Nature and man were not to be denied, but the sailors had their orders; not only were they seamen but also deputized keepers of the social order. They locked another gate to stop the flow of humanity, and three Irish girls—Kate Gilnaugh, Kate Mullins, and Kate Murphy—reached the gate as a seaman clicked the lock. They pleaded with the man to let them through, but the seaman wouldn't budge until Jim Farrell, a large Irishman, slammed against the gate and roared out, "Good God Man, open the gate and let the girls through!"[7] To the girls' amazement the seaman opened the gate and they proceeded on to the upper decks.

But many of the third-class passengers could not find their way out. They simply milled around in bottlenecked areas around the forward deck or the bottom of the E Deck staircase. No one was telling them what to do or where to go. Any organization for abandoning the ship was reserved for the first-class passengers. Many returned incredibly to their cabins, effectively giving up. Many turned to prayer. Third-class passenger Gus Cohen saw many in the third-class dining room with heads bowed and rosaries clutched. At this point, all that was left to them was prayer.

Lightoller was still hard at work when he saw the engineers come trooping up from below. These were the men who had kept the lights on by keeping the dynamos running. They emerged into a world of people running from the ocean that was creeping up over the ship with no apparent lifeboats left. "They had all loyally stuck to their guns, long after they could be of any material assistance,"[8] Charles Lightoller later wrote. "Up to that time they had known little of what was going on, and it was surely a bleak and hopeless spectacle that met their eyes. Empty falls hanging loosely from every davit head, and there was not a solitary hope for any one of them."[9] Lightoller noted that out of the thirty-five engineers, many of whom he knew from other ships, not one survived.

It is at this point that the collapsible boats had become the only escape for the more than fifteen hundred people remaining. The careful order so persevered in many histories of the *Titanic* broke down. When one of the boats was readied, someone shouted, "There are men in that boat." Lightoller reacted violently. "I jumped in … and they hopped out mighty quickly and I encouraged them verbally also by vigorously flashing my revolver. … I had the satisfaction of seeing them tumbling head over heels onto the deck, preferring the uncertain safety of the deck to the cold lead."[10]

Lightoller noted later that the revolver was not even loaded. He then took over the hooking of Engelhardt B on the davits and then swung it out for lowering. He stood partially in the boat helping the women over the high rail when First Officer Murdoch walked over and saw no seamen were available. "You go with her, Lightoller." The second officer who had long since taken off his coat and was sweating profusely in the cold night looked at his superior officer and shook his head. "Not damn likely."[11] Lightoller jumped back on board, and as the boat lowered two men jumped in from a deck below.

Meanwhile, Jack Phillips was still sending out his CQD and SOS to anyone who would respond. It was 1:15, and he was beginning to wonder if anyone could reach them in time. The *Baltic, Virginian, Birma, Parisian, Cincinnati, Prinz Friedrich Wilhelm, Prinz Albert,* and *Caronia* all were heading toward the *Titanic* but were too far away to make a difference. Only the *Carpathia* and *Mount Temple* had a shot at getting to the *Titanic* in time or least rescuing the people in the lifeboats. The only good news was that *Carpathia* was coming fast, but she was fifty-eight miles away, and that meant at least four hours. Phillips had not heard from *Mount Temple* directly, but the message was relayed that she was less than fifty miles away and headed for his position. He needed someone closer, a ship that might be able to reach them before *Titanic* slipped under the icy water. One problem was that many of the ships had a hard time believing an unsinkable ship like the *Titanic* was going to sink. Even the *Olympic* operator had a difficult time understanding, asking if they were going to meet them. It was the *Frankfurt* that put him over the edge. "Are there any ships around you already?"[12] Phillips ignored him. "The *Frankfurt* wishes to know what the matter is. We are ten hours away." Harold Bride then saw Phillips jump up and pull the headphones off and shout, "The damn fool!" He then responded, tapping furiously, "You are jamming my equipment! Stand by and keep out!"[13]

Captain Smith was feeling the strain as well by now. He had been dropping in every fifteen minutes to see if any other ship had responded that was closer than the *Carpathia*. He could not believe that the light off the bow that he had the lifeboats rowing toward, despite Lightoller Morsing and the rockets exploding overhead, had not responded. It was inconceivable to him they should sink with a ship that close and get no assistance. It made no sense. Phillips stayed bent over his key, sending out his signals, hoping some ship might turn on their ears and come to their aid, but he realized by the tilt of the ship that they were running out of time. Phillips decided to take a break and see for himself what was going on outside the insulated wireless room where they could hear very little.

Bride took over and Phillips opened the door and walked out into the cold air. He had forgotten how warm the wireless room was and realized this icy cold was coming for all of them when the ship finally sank. He walked and saw men rushing toward a few boats with *Titanic* officers holding them back. Phillips saw the sea sweeping over the bow of the ship, "over the foredeck, washing past the foot of the foremast, swirling around the winches and cranes, flooding into the forward well deck."[14] It was shocking, and he knew then, even with the *Olympic, Carpathia, Baltic, Virginian, Birma,* and *Mount Temple* all rushing toward the *Titanic,* that no ship would get there in time. He turned and saw a light on the horizon—the ship that Captain Smith had spoken of. It was eerie that this ship should be this close and had not responded to his CQD or SOS. He must not be equipped, or worse, the operator had gone to bed. Surely, surely, they saw the rockets and the Morse signaling. Surely someone would wake the operator and tell him to put on his headphones.

Phillips walked back into the warmth of the wireless room and breathed in the warm electric scent of the transmitter. Here was science and outside was chaos. This little room perched atop the sinking behemoth was more powerful right now than the 50,000 horsepower of the giant reciprocating steam turbine engines. The wireless room was like a flea on a dog and yet this flea was their last best hope against oblivion. This warm little center of the universe had the power to bring help from beyond the frozen darkness. Phillips relieved Bride and sat back down. When Bride asked him what it looked like outside, Phillips shook his head.

"Things look very queer outside, very queer indeed."[15]

He thought of the light again on the horizon and began tapping. "CQD SOS CQD SOS . . . CQD SOS. . . ." It was their last best hope, and he wouldn't stop until the generators died. Or he did.

CHAPTER EIGHTEEN

A Brilliantly Lit Ship

80 Minutes

PASSENGER E. W. ZURICH FELT LIKE A SCHOOLBOY. HE AND TWO OTHER companions had disobeyed orders to not go on the deck of the *Mount Temple*. He pushed against the heavy metal door, freezing to the touch, and heard the squeal of the frozen hinges. The air was so cold it froze the inside of his nose. Above he saw a million bright stars. Zurich walked to the railing and there in the night was a brilliantly lit ship.

To this day there remains confusion as to how far the *Mount Temple* was from the *Titanic* when she received her distress call. Some say fifty miles. Some say forty miles. The passengers and crew put her much closer, so close that Captain James Henry Moore would come face to face with a choice of entering the ice field while the *Titanic* was still afloat and in range of rescue. The *Mount Temple* had received the first distress call at 12:15 a.m. Captain Moore had immediately put about and headed for the position given by the *Titanic*. He had 1,461 passengers on board his ship and came upon an ice field reportedly at 2 a.m. The *Titanic* was still afloat, putting off people in boats and shooting up rockets. Like Stanley Lord of the *Californian*, Captain Moore stopped his engines when he approached the ice field that had trapped the *Titanic*. Like Stanley Lord he was confronted with a choice, and like Stanley Lord his choice would be observed by crew and passengers.

The passengers had been forbidden to go topside while the crew readied the lifeboats and swung them out. But when the *Mount Tem-*

ple stopped in front of a vast ice field, and the vibration of the engines ceased, passenger Zurich opened the door onto the boat deck. He knew of the 12:15 distress call from the *Titanic* that had resulted in the altering of their course and of the lifeboats being swung out from their davits. He also knew of the order given that no passengers were to be allowed outside on the upper decks, but Zurich and two other passengers were curious.

The ship had stopped, and the frigid air watered Zurich's eyes as he stared out into a vast ice field. Icebergs that looked like white polar bears dotted the ocean. His breath steamed as he stepped toward the railing with his two companions. The smell was that of an icehouse. Out in the darkness he could see a glittering ship with tall funnels and masts. A reporter who later interviewed the *Mount Temple* passenger wrote,

> *He is fairly positive that they saw the masts of the* Titanic, *and he says he is not ready to accept the assertion that their ship was at least forty miles from the wrecked liner at the time. At any rate he thinks the* Mount Temple *might have reached the spot before the* Titanic *sank and this supposition, he says, seemed to have been entertained by others on board.*[1]

The *Buffalo Commercial* led with a front-page headline on April 25, 1912, "Turned Away from the Sinking Ship, Passengers and Members of Crew of Steamer *Mount Temple* Say They Saw *Titanic's* Signals."[2] The *Times-Republican* had multiple headlines: "*Titanic* Call Ignored! *Mount Temple* Refused to Enter Ice Fields . . . Light of Wrecked Ship in Plain Sight."[3]

Captain Moore would have been fine if crew and passengers had stayed below, but instead they became witnesses to a moment better left to the darkness. In interviews, Zurich described the moment the *Mount Temple* saw the *Titanic* and turned back:

> *According to Mr. Zurich, passengers on board the* Mount Temple *heard of the* Titanic's *distress at 12:15 o'clock Monday morning when a wireless call for help was caught. Captain Moore changed his vessel's*

course at once and headed for the Titanic, *lifeboats being swung from davits meanwhile, and other preparations being made for lending assistance. The Northern course was not held long, however, says Mr. Zurich, because a great field of ice loomed up ahead. It was reported among crew and passengers, according to Mr. Zurich, that Captain Moore made no further efforts to penetrate the flow, asserting that he could not afford to take the risk of endangering the 2,000 passengers on board his ship.*[4]

If Captain Moore had turned back from ice on a normal crossing of the North Atlantic, it would have been regarded as prudent, but stopping in sight of *Titanic* made this deplorable. Another paper, the *Elmira Star Gazette*, led with the headlines on April 25, 1912:

Mount Temple *Passengers Saw Signals from* Titanic . . . Dr. *Quitzrau Positive He Beheld Masts of Foundering* Titanic, *While Members of Crew Allege They Heard Officers Discussing Distress Messages (rockets) Which They Saw . . . State Ship Deliberately Declined to Give Aid to People On Board the* Titanic.[5]

Dr. Quitzrau was one of the passengers who snuck up to the deck and saw the *Titanic* as well. He later gave an interview to the *Toronto Star* about a conversation that placed *Mount Temple* only forty miles away when she received the distress signal. "The statement of Dr. Quitzrau to the effect that passengers and crew believed they could see the lights of the unfortunate *Titanic* is borne out by Mr. Zurich."[6] Quitzrau then filed a deposition that was sent to the inquiry in Washington on the sinking of the *Titanic*:

About midnight Sunday, April 14, New York time, he was awakened by the sudden stopping of the engines, that he immediately went to the cabin at which were already gathered several stewards and passengers who informed him that word had been received by wireless from the Titanic *that the* Titanic *had struck an iceberg and was calling for help. Orders were immediately given, and the* Mount Temple *course changed, heading straight for* Titanic. *About 3 o'clock New York time, 2 o'clock ship's time, the* Titanic *was sighted by some of the officers and crew, that as soon as* Titanic *was seen, all on the* Mount Temple *were put on alert and the engines stopped and the boat lay dead for about two hours.*[7]

Quitzrau then stated that at daybreak, "The engines were started, and the *Mount Temple* circled *Titanic*'s position, the officers insisting that this be done."[8] Another passenger watching from the deck was W. H. Kenerworst, who later told the *Indianapolis Star* that the *Mount Temple* "was within five miles of the *Titanic* half an hour before the liner went down."[9] Kenerworst went on and described what he saw that night and

declared he saw the lights of the Titanic *and that although Captain Moore of the* Mount Temple *had received wireless messages that the White Star liner was sinking and that the women and children had been put off in boats, he hove to his ship in spite of the entreaties of his officers that he rush to the aid of the* Titanic.[10]

It gets worse for Captain Moore, according to Kenerworst's account, which asserted that his own officers, "clothed in high boots and heavy overcoats, ready to lower *Mount Temple*'s boats, had urged Captain Moore to make an attempt to reach the *Titanic* but that Captain Moore had replied it was too dangerous and he would not risk the lives of his own passengers."[11]

The crew of the *Mount Temple* were also watching that night and saw more than just the ship; they saw the rockets. "Sailors, firemen, and others declare that they sat on the deck for hours and watched the *Titanic* sending up rockets and burning red and blue light, until the *Mount Temple* steamed so far away that these signals were lost."[12] It gets worse still. A sailor on watch on the bridge heard Third Officer Notley "tell the captain of the distress message (rockets), and that instead of the steamer heading directly to the wreck, she steamed away on her own course, so that the lights were soon lost."[13] An oiler named Pickard on duty at the time asked the men to "keep her fired up to the limit, as it was a case of life or death."[14] The *Buffalo Commercial* interviewed another sailor from the engine room who added,

> When his watch was over he went on deck and, with many others, passengers and crew, leaned over the rail and saw the almost steady stream of rockets being sent up by the Titanic. He adds that in spite of the cold of the night he remained on deck until almost 2 o'clock watching until the signals were lost in the distance. His version of the affair is that all the time the Titanic was in distress the Mount Temple was only between five and ten miles from the place.[15]

The life and death of the ice field would soon be behind the ship, but not before others saw the rockets as well. Captain Moore claimed to have seen a tramp steamer that night before hitting the ice field. "I met a small schooner, a small craft, and had to get out of the way," he said. "The schooner's lights seemed to go out later. It was between me and the *Titanic* on our port bow. . . . I was going due east, and the schooner was coming from the direction of the *Titanic*."[16] The existence of the tramp steamer would never be proved and would become the elusive mystery, or ghost, ship. Moore never denied he stopped for the ice. He said,

> We received a wireless message after midnight Sunday from the Titanic *stating that she had struck an iceberg and to come at once. We turned about at 12:30 and steamed back to the position given us, arriving there at 4:30. We encountered so much ice, however, that we stopped until daylight.*"[17]

Moore estimated he was fourteen miles from the *Titanic* when he stopped for the ice and decided to go no farther. The *Mount Temple* had intercepted messages between the *Carpathia* and *Titanic* earlier. *Carpathia*: "Do you want any special ship to stand by?" *Titanic*: "We want all we can get."[18] Clearly, Moore knew how desperate *Titanic* was for assistance. He would later deny that the *Titanic* was visible to crew and passengers, stating, "I can solemnly swear that I saw no signal lights from the *Titanic*. . . . They are mistaken. There were no passengers on deck at that time. I was on the bridge myself."[19] But he did not know about Zurich and his companions or of his own crew who were observing the ship shooting off rockets. So, what really happened on the *Mount Temple*? What Zurich and his two companions and the crew really saw that frigid night was the *Titanic* in her death throes. Captain Moore steadfastly denied this, but, like Stanley Lord, he made a decision about the danger of entering the ice, later revealing it in an interview: "I had at least 1,600 people on board my steamer, and it would have been very foolish for me to try and force my way through five miles of ice floe, for undoubtedly I would have met with the same disaster as did the *Titanic*."

Then Captain Moore reveals his deepest fear: "The ice was so thick that it would have cut through the iron plates on the ship like paper, so I decided that it was useless. I would have been very thankful if I could have been of assistance, but God knows I did all that I could."[20] Captain Moore did do all he could, except enter the ice field where the RMS *Titanic* was sinking into oblivion. Moore had found himself just outside of the ice field with the small lit image of the *Titanic* in the distance, shooting up arcs of fire-trailing rockets that exploded like small electric puffs in the brilliantly clear night. Moore had a choice either to stop or to risk his own life and others to rescue a ship sinking directly in front of him. The problem for the captain was that this hidden moment was witnessed by passengers and crew, who became so outraged with Moore's decision to sit and do nothing that they considered a mutiny. This is the dirty secret that has nipped at the corners of the *Titanic* mythology of a ship beyond help and a heroic Edwardian last stand replete with chivalrous moments. It is impossible to believe that the crew, the passengers, and the wireless operator all fabricated stories to different papers with events that mirrored each other. Moore's own words belie his decision not to enter the ice field. The truth is the *Titanic* could have been rescued by the *Mount Temple*. Captain Moore simply didn't have the courage to enter the ice field to do it. All the passengers and crew could do was watch as a ship carrying more than 2,200 people sank in front of them. Moore might have slinked off into the darkness of history if it were not for the *Carpathia* and Captain Rostron, who showed the courage so lacking in others that night.

CHAPTER NINETEEN

Racing through the Night

70 Minutes

ON BOARD THE *CARPATHIA*, MRS. ANNIE CRAIN WOKE UP AT 1 A.M. TO the smell of fresh-brewed coffee. Her cabin was cold, and the ship was vibrating. She could not help but feel something ominous was afoot. She had sailed on the *Carpathia* several times and knew the rhythms of transatlantic voyages. At night, the cabins were usually toasty, and the vibration from the engine and the screws created a soft lulling background that she felt aided in her sleep. Never had she smelled coffee before in the middle of the night, and never had her room been so cold. Just down the corridor, Mrs. Ogden had shaken her husband awake. She had woken up in her bed cold with the vibration of the ship shaking the water glass next to her bed. The Ogdens too had been on the *Carpathia* before, and never had they experienced the sensation Mrs. Ogden felt now. The ship was heaving up and down, and Mr. Ogden looked out the porthole, deducing that the ship was racing through the night at a speed he didn't think she was capable of. He then became aware of heavy thumps overhead and heard the squeal of davits and the unmistakable sounds of pulleys. Ogden knew then the lifeboats were being lowered or at least uncovered, and this got him out of his bed and into the hallway.

He looked down the corridor, breathing in the scent of coffee and seeing "stewards and stewardesses carrying blankets and mattresses."[1] This was too much. He went into his room to dress and get topside as quickly as possible. Across the ship on A Deck, Howard M. Chapin woke much the same way as Ogden. He woke to hammering above him and remembered that a cleat used to tie off the falls of a lifeboat was located over his stateroom. He could hear the line being dragged across the deck above him and knew then the lifeboat cover had been unfurled. That meant only one thing: The *Carpathia* was in trouble. But more than that, she was going hell to leather somewhere. The entire ship was vibrating so much he felt a tickling down in his stomach. The glasses were rattling next to the washbasin with the decks and bulkheads literally humming with momentum. Chapin had been on enough ships to know this was highly unusual. The *Carpathia* was being driven to go faster than she was designed for, and only a real emergency would necessitate this kind of speed. He would go find out what was going on.

These same moments began occurring throughout the ship. Captain Rostron's admonishment to keep the news of the *Titanic* from the passengers and passengers in their cabins was not working out. You could not divert every ounce of steam into the engines and have every stoker firing up the boilers with as much coal as possible and go unnoticed. You could not drive your ship beyond the speed she was designed for, while zigzagging between icebergs while men strained to see into the night and remain veiled in secrecy. You could not have the entire crew awake and feed them coffee while they set up triage stations for frozen passengers that might have spent time in the North Atlantic and not expect the passengers to wake up and come looking for answers. Passengers huddled around the upper decks and in the corridors, asking stewards why their rooms were so cold and why the ship was racing through the night. The stewards could not answer the questions as they carried mattresses and blankets because they didn't know the answer either. Ogden believed the ship might be in danger of sinking and speculated she was making a run for shore. Rumors rifled through the ship, and while it was not panic setting in among the passengers, there was a growing buzz of fear as more people emerged from their cold staterooms.

Chief Steward Henry Hughes knew about *Titanic*. It had been an hour since Cottam had picked up the CQD and Hughes now believed it was time to tell the crew and felt it would galvanize them into higher performance of their duties. He had the entire crew of stewards and stewardesses assemble in the main dining room. At 1:15 everyone was there. He told them about the *Titanic* hitting an iceberg and how they were in a race to rescue her passengers before the great ship sank. He reinforced that they were the *Titanic*'s best hope and it was up to the *Carpathia* to save as many lives as possible. He paused then and spoke solemnly, "Every man to his post and let him do his duty like a true Englishman. If the situation calls for it, let us add another glorious page to British history."[2]

Hughes was right. The stewards now set about their work like men possessed and were determined to be ready for any eventuality. But the passengers knew none of this, and when Louis Ogden emerged from his room fully dressed, he ran into the ship's surgeon, Dr. McGee. Ogden asked the doctor what was going on and was told to return to his room. Ogden was not to be put off and demanded to know the situation that had the ship pounding across the Atlantic with the crew up and carrying mattresses and blankets. McGee sighed and then lowered his voice. "An accident, but not to our ship. Now please go back to your room."[3]

Ogden did go back to his room, where he and his wife had a quick panicked conversation, coming to the conclusion the *Carpathia* had caught fire and it was time to head for the open decks before panic set in. Ogden then left his cabin again. Like a stealthy commando, he slipped into a side door and emerged on the upper deck, where a wall of cold air slapped him awake. He saw a quartermaster he had known from previous voyages on the ship. Ogden called to him from the shadows and demanded to know what was going on. The quartermaster lowered his voice and quickly told him that *Titanic* had struck an iceberg and they were going to her aid. Ogden scuffed, not willing to give up his conviction the ship was on fire. "You'll have to do better than that! We are on the Southern route, and *Titanic* is on the Northern!"[4] The quartermaster had had enough of this man who should not have been out of his cabin in the first place. "We're going north like hell," he snapped. "Now get back to your cabin!"[5]

The quartermaster left. Ogden paused, then went to his cabin, where he had another conversation with Mrs. Ogden, who wasn't buying the quartermaster's story either. It was preposterous. Everyone knew the

Titanic was unsinkable. Mrs. Ogden dressed in her warmest clothes, and they both headed topside and reached some other passengers gathered on the promenade deck. They quickly compared notes: a ship speeding through the night, lack of steam in their rooms, a crew powered by black coffee to work through the night. And now they found they had all been told the same incredible story—that the largest, most luxurious ship in the world, the unsinkable *Titanic*, was sinking, and the *Carpathia* was going full steam to try and save her passengers. They collectively shook their heads. It simply was not possible.

CHAPTER TWENTY

I Believe She's Gone, Hardy

50 Minutes

JACK THAYER AND HIS FRIEND LONG WERE TRYING TO DECIDE WHAT TO do. They watched Second Officer Lightoller and crew members as they worked to free up an Engelhardt collapsible boat on top of the officers' quarters. They considered sliding down the empty lifeboat falls dangling over the side of *Titanic* like tendrils of the lifeboats that had been their last chance. As Thayer later wrote, the falls "were swinging free all the way to the water's edge, with the idea of sliding down and swimming out to the partially filled boats lying in the distance."[1] He reasoned they would be away from the "crowd and away from the suction of the ship when she finally went down. We were still 50 or 60 feet above the water. We could not just jump, for we might hit wreckage or a steamer chair and be knocked unconscious."[2]

Long talked Thayer out of it and they continued to walk the deck, contemplating the thought of drowning in the icy Atlantic. Most of the sixteen wooden lifeboats were now gone, and the passengers lucky enough to have made it into the boats were now watching an incredible scene. Thayer gives one of the few snapshots of what the scene on board the *Titanic* was like as she edged deeper into the Atlantic:

> On deck, the exhaust steam was still roaring, the lights were still strong. The band, with life preservers on, was still playing. . . . Our own situation was too pressing, the scene too kaleidoscopic for me to

retain any detailed picture of individual behavior. I did see one man come through the door out into the deck with a full bottle of Gordon's gin. He put it to his mouth and practically drained it.[3]

The lifeboats rowed away from the *Titanic*, some smoothly, some going in circles with people rowing who had never rowed a boat in their life. The glassy sea only made the scene more surreal as if they were in a giant water tank and a scene was being filmed involving a giant liner slipping into the ocean. There was no way not to watch the *Titanic* now from the boats. She towered over the little armada with her looming funnels and masts and the brightly lit promenade decks and all the portholes that were now circles of yellow light slowly being submerged into the green, translucent water. People were lining the rails as if waving goodbye on a long voyage, and the band still played ragtime with the eerie music sailing out into the fathomless void. Except for the fact that the giant skyscraper of a ship was now head down into the sea, it was still impossible to believe that this glittering monolith, this city of lights, was actually going to disappear beneath the people in the lifeboats.

Lawrence Beesley, in Boat 13, in decorous Victorian prose tried to convey the scene he witnessed:

Here again was something quite new to us; there was not a breath of wind to blow keenly around us as we stood in the boat. . . . The sea slipped smoothly under the boat, and I think we never heard it lapping the sides, so oily in appearance was the water. . . . The mere bulk of the ship viewed from the sea below was an awe-inspiring sight.[4]

Some of the boats had been told to stand by while others had been instructed to row toward the steamer whose yellow light lay tantalizingly close. Captain Smith had instructed those in Boat 8 to row toward the steamer and land her passengers and then come back for more. That was how close the ship seemed to be. He then asked Quartermaster Rowe to try Morsing the ship again. "Call that ship up," he commanded. "And when she replies, tell her, 'We are the *Titanic* sinking, please have your boats ready.'"[5]

Fourth Officer Joseph Boxhall had been trying for some time to Morse the ship, but Rowe thought he might be successful and instantly saw another light on the horizon that he pointed out to Captain Smith. Smith looked through his binoculars and told the quartermaster it was a planet but to keep trying and keep watching. There were many ships headed for *Titanic* now and who knows who might have heard their CQD and SOS signals. The hope was that a ship close enough was on the way and had not responded to the wireless call Phillips had been putting out nonstop. From his lifeboat, Beesley saw a beauty that was not evident onboard the *Titanic*. "We saw it all. All sense of the beauty of the night, the beauty of the ship's lines, and the beauty of her lights, and all these things . . . the awful angle made by the level of the sea with the rows of porthole lights along her side in dotted lines, row above row."[6]

But there was no beauty on the *Titanic* now. Only terror. The steam blasting though the shipboard funnels had stopped, but a different pandemonium was taking place, a creeping realization that most of the boats were gone and that most of the people still on the *Titanic* would die. There had been no general announcement that there were far too few boats for the 2,200-some passengers. If there had been, it would have sparked an early panic with the knowledge that even if every boat were loaded to capacity well over a thousand people would still die when the ship sunk. So now it was simple math. The people on board looked around and saw no boats or just a few and then they saw the many souls still on the ship. The realization that there was no plan to save them, and worse, there was no ship that could reach them in time, must have sent many into a state of shock.

The ship listed more to port from the weight of passengers crowding the rails, looking in vain for any remaining lifeboats. Chief Officer Wilde realized the danger, and at twenty minutes to 2 a.m. he shouted at the people crowding the rails, "Everyone on the starboard side to straighten her up!"[7] It was amazing that a ship the size of the *Titanic* could be affected by the weight of passengers, but the massive tonnage of water in the bow now made the ship unstable, and it did come back to center when the crew and passengers went back across

The saga of Boat 4 was an ongoing story. An hour before, it had been lowered to A Deck to be filled from there, but the windows were all sealed. Then someone saw the sounding spar sticking out below it. Seaman Sam Parks and storekeeper Jack Foley went to look for an axe to chop it away and Second Officer Lightoller went on with other boats. The passengers waiting to go in Boat 4 were a who's who of the Gilded Age: "The Astors, Wideners, Thayers, Carters, and Ryersons were sticking pretty much together." The husbands showed up while the women and children waited silently as the spar was chopped off and the windows on the deck below were opened. The waiting elites were ordered up to the boat deck and then back down again. Mrs. Thayer looked at a steward: "Tell us where to go, and we will follow! You ordered us up here, and now you're sending us back!"[8]

Lightoller returned and took charge. He put one foot on an open window and one on the boat and then had chairs put by the rail for the women and children to use. When Mrs. Astor entered the boat, her husband John Astor asked if he might go as well since she was in a "delicate condition."[9] Lightoller looked at the wealthy man and shook his head. "No sir . . . no men are allowed in these boats until the women are loaded first."[10] Astor then asked the number of the boat. Lightoller later said he thought he wanted the boat number to lodge a complaint. But money was doing nobody any good on the *Titanic* now. Rich and poor were staring at the same fate. The Ryersons then boarded Boat 4. Mr. Ryerson grave his life jacket to their French maid, Victorine. Mrs. Ryerson led their thirteen-year-old son to the window and Lightoller held up his hand, "That boy can't go."[11] Mrs. Ryerson stood up to the seaman who had seen his share of wrecks. "Of course, that boy goes with his mother—he is only thirteen."[12] Lightoller let him go, grumbling, "No more boys."[13] Finally, at 1:55 a.m., Boat 4 lowered away, and Mrs. Ryerson, watching from the lifeboat as it lowered, saw the water gushing into the ports on C Deck and rushing around the furniture in the first-class suites. She glanced up and saw her husband looking down, standing by Widener in profound silence.

Now there was just Collapsible D, which had been put into the davits of Boat 2 and was the last to load. The lights were beginning to dim and from below came the sound of plates and glassware crashing as the ship listed farther into the sea. Lightoller had the crew lock arms around the boat which had room for only forty-seven people. There were roughly still sixteen hundred on the ship and Lightoller knew it could get ugly. Famously, two babies were passed through the men guarding the boat by a man who identified himself as Mr. Hoffman. It was later revealed that his real name was Navratil and he had kidnapped the babies from his estranged wife. A famous theater producer, Henry B. Harris, ushered his wife to the ring of men and was stopped and told he could not join her. "Yes, I know. I will stay,"[14] he sighed. Then Major Butt escorted five unprotected ladies to the line along with Mrs. John Murray Brown and Miss Edith Evans. The women passed and Butt stayed, even though he was the special assistant to President Taft. The boat was full and starting to go down when Evans turned. "You go first. You have children at home."[15] She helped Mrs. Brown into the boat as the boat was lowered away and then watched.

Hugh Woolner and Bjornstrom Stefansson were by the rail when they went onto the promenade deck. The sea flooded out of the open door and quickly went up to their knees as they jumped on the railing. They saw Boat D lowering down the side of the ship less than ten feet away. The last boat of salvation. "Let's make a jump for it,"[16] shouted Woolner. "There's plenty of room in her bow!"[17] Stefansson jumped over the side and took a flying leap into the front of the lifeboat and landed upside down. Then came another body out of the darkness, and Woolner crashed down in the front of him with half his body hanging out. The two men sat up with the women and children staring at them. It did not matter, they had escaped. Boat 2 was one of the last to go at 1:45, and a steward named Johnson yelled for a knife to cut the falls. Seaman McAuliffe tossed his down from the boat deck and shouted, "Remember me at Southampton and give it back to me."[18] There were few in the crew now who believed they would return to Southampton, and First Officer Murdoch confirmed the worst when, standing next to Chief Steward Hardy of second class, he murmured, "I believe she's gone, Hardy."[19]

The ironclad rule of woman and children first was breaking down as human beings realized protocol, chivalry, and honor meant nothing if you were dead. People did anything they could now to get into the few lowering boats. A man leapt ten feet off the railing and landed in a boat dangling over the side. Seamen on the lower decks tried to pull him out, but he wiggled away and made his getaway from the moribund ship.

Fifteen-year-old Daniel Buckley had made his way through a labyrinth of stairs and corridors from steerage to finally emerge onto the boat deck, where he quickly realized there were few boats left. He and some other third-class passengers crowded into a remaining boat, crying in broken English as seamen pulled them out. Buckley grabbed a woman's shawl and put it over his head assisted by Mrs. Astor, and when the boat descended Buckley kept his head down and was able to escape the ship. But Fifth Officer Lowe caught a boy who was hiding under a seat in Boat 14, and when he would not leave, Lowe drew his pistol and the boy begged and cried to let him remain. Lowe was not going to shoot the boy but told him simply to be a man and shamed him into getting out, even as Mrs. Charlotte Collyer and other women cried and begged him to let the boy stay. Mrs. Collyer's eight-year-old daughter tugged on Lowe's sleeve, "Oh, Mr. Man, don't shoot the poor man!"[20] Officer Lowe nodded to her and moved on with the boy lying face down near a coil of rope.

The chivalry was all but gone along with the nineteenth-century patina of the Gilded Age painted over the rot beneath. This rot was the reality of human existence where the struggle to survive trumps all. The *Titanic* mythology of the orderly departure from this life, created by a lack of real news, covered up some awful truths about human failings that night. The lower classes were now in control of the ship. They had been kept safely down in steerage with blocked corridors and stairs, but the people now overwhelmed the sinking ship, running wildly from one side to another while seamen fought for order and then gave up.

A mob of men rushed Boat 14 until a seaman swung the boat's tiller, clubbing several of them to the ground. Lowe saw the altercation and shouted at the men, "If anyone else tries that, this is what they'll get!"[21] Lowe then fired his pistol three times along the side of the ship. The shots mark a clear divide in the saga of the sinking. The syrup of patriarchal white Christian male dominance that has been poured all over every history of the *Titanic* is now thinning out under the reality. Why would the *Titanic* not devolve into a riot? People were now fighting for their lives. Officer Murdoch blocked a rush on Boat 15, yelling at the men, "Stand back! Stand back! It's women first!"[22]

Captain Smith on the bridge undoubtedly saw the fight break out around Collapsible C as a group of third-class passengers and stewards charged the boat. The significance here is that order among the stewards and crew was breaking down. The stewards were just as human as anyone else and they realized their only chance was to get into anything that would float. Pursuer McElroy was by the collapsible and pulled out his revolver, firing twice into the air. Murdoch shouted, "Get out of this! Clear out of this!"[23] The mob jumped back as Murdoch and two first-class passengers pulled out the interlopers.

The pathos in this scene is heavy. As Captain Smith looks on, the mob has broken from the bowels of the ship and now infected even the stewards, who rush the ramparts where a single pursuer holds them back while the first-class passengers, the ruling class, drag the lesser humans out of the boat. This line of order was now moving much like rioters in a city where the police chase them and try to anticipate their next move. The mob moved on and descended on Boat 2, where third-class passengers and crew sat huddled waiting to be lowered away. That was when Second Officer Lightoller pulled out his revolver, shouting, "Get out of there you damned cowards! I'd like to see every one of you overboard!"[24] The men scrambled out of the boat, having heard the shots fired by Officer Lowe, and Lightoller looked perfectly capable of shooting them. He turned the boat over to Fourth Officer Boxhall, who loaded the boat quickly with twenty-five women, one male third-class passenger, and three crewman and lowered away.

Seventeen-year-old Jack Thayer and his friend Long watched the drama unfold from the starboard side of the boat. "We were a mass of hopeless, dazed humanity, attempting, as the almighty and nature made us, to keep our final breath until the last possible moment."[25] They saw no real escape. "At times we were just thoughtful and quiet, but the noise around us did not stop."[26] Thayer and Long exchanged messages for their families if only one of them should survive and then went back to contemplating their fate.

So many thoughts passed so quickly through my mind! I thought of all the good times I had had, and of all the future pleasures I would never enjoy; of my father and mother; of my sisters and brother. I looked at myself as though from some far-off place. I sincerely pitied myself. . . . We still had a chance, if only we could keep away from the crowd and the sinking of the ship.[27]

They watched the water creeping up the deck. Thayer noted it was right up to the bridge now and estimated "there must have been 60 feet of it on top of the bow."[28] This is incredible. At 2 a.m., the bow of RMS *Titanic* was down sixty feet into the Atlantic Ocean. Thayer and Long watched the people flee the encroaching sea. "As the water gained headway along the deck, the crowd gradually moved with it, always pushing, toward the floating stern."[29] Thayer knew his time was running out and he would soon be in the freezing water creeping toward him like a creature of the doomed.

CHAPTER TWENTY-ONE

Speed Is of the Essence

40 Minutes

VICE PRESIDENT PHILIP FRANKLIN OF THE INTERNATIONAL MERCAN-
tile Marine was sound asleep when his phone rang in his Upper East
Side home in New York. The 1912 phone bell was rich and sounded like
a trolley car in the hallway. He looked at the clock next to his bed. It was
2 a.m. He picked up the receiver and held it to his ear. Charles E. Crane
of the Associated Press squawked loudly and told Franklin they were
getting reports that the *Titanic* had struck an iceberg and was sinking
and had sent out calls for assistance. Franklin blinked in disbelief, feeling
a sudden chill. He was standing in his pajamas and bare feet and could
not believe the words he was hearing. He asked Crane for the source of
his information. He said it had come from the *Virginian*, which had communicated with Montreal. This made no sense, but the report had come
from an Allan Line employee who informed a Montreal reporter that
the *Titanic* had sent out a CQD the *Virginian* intercepted. The *Montreal
Gazette* had a content-sharing arrangement with the *New York Times* and
had given the information to managing editor Carr Van Anda, who was
about to have the scoop of the decade.

Franklin suddenly woke up and started yelling at the reporter. Crane
later said Franklin shouted, "Stuff and nonsense it was to him. The
Titanic would shatter an iceberg into a frappe. He hung up with, 'Don't
call me again.'"[1] Franklin put the receiver back on the hook and found
his hands shaking. Reporter Crane was not to be put off. He wrote a

story anyway that said the *Titanic* was sinking with a large loss of life. Franklin picked the phone back up and called the White Star dock. An official told him they had no information but had received calls from several reporters. He then called the Associated Press office and asked them to hold off on the story until he could confirm it. It was too late; the story was already out. The story was getting out, but reporter Crane's story would not see the light of day. Crane was asleep when he got a call from the AP copydesk that said the White Star Line was denying the story, and they had decided not to run it. The fact was, newspapers on both sides of the Atlantic were going to press with bits and pieces, but no one had the full story, yet.

Franklin then contacted the line's Montreal office to track down the report from the *Virginian*. He contacted the head of IMM steamship department who then was instructed to contact the captain of the *Olympic*. "I do not want to alarm the captain of the *Olympic*," Franklin said. "So, all I asked in the telegram was, 'Can you get me the position of the *Titanic*? Wire us immediately her position.'"[2] Now Franklin had to wait to hear back from the *Titanic*'s sister ship. The *Olympic* had some of the most powerful wireless equipment:

> *Its generating plant consisted of a 5-kw motor generator set, yielding current at 300 volts 60 cycles . . . the guaranteed working range of the equipment was 250 miles under any atmospheric conditions, but actual communication could be kept to about 400 miles, while at night the range was 2,000 miles.*[3]

There were four parallel antenna wires strung across the funnels of the *Olympic* for reception. *Olympic* could overwhelm other ships and communicate directly with the *Titanic*. Still, Franklin's inquiry would have to be relayed first to Sable Island and then to the ship. All Franklin could do was wait.

On the *Olympic* Captain Haddock had every boiler lit and every stoker pounding in as much coal as possible. She was five hundred miles away, but Haddock intended to close the gap even if he had to freeze his passengers by diverting steam to the engines. He was also trying to keep the *Titanic's* sinking away from his passengers, but he was being undone by his incredibly famous passenger who had had overseen the design and the building of the 1893 Chicago World's Fair. Daniel Burnham and his wife, Margaret, on board the *Olympic*, had stopped in Washington for a meeting of the Lincoln Memorial Commission before traveling to New York to catch the *Olympic* on the 13th. It was there he wrote a letter to his friend and partner on the World's Fair project, Frank Millet, who was traveling on the *Titanic* with Major Archibald Butt, military aide to President Taft.

The letter was to be given to Millet when he arrived in New York.

Dear Frank, my wife and I sail tomorrow on the Olympic *crossing you at sea. I am writing this letter to be handed to you on landing . . . at the end of the council a vote for designer was about to be taken, but the President deferred until sometime next week. I am writing to the President now asking that it be when you can be present, for I feel that the decision is going to be a vital one, settling for a long time the status of the fine arts in this country.*[4]

Burnham knew the *Olympic* and *Titanic* would be passing in mid-ocean, so he wrote a message for Millet and had it delivered to the Marconi office for transmission. The steward came back with the message and explained it could not be sent. Burnham, the most famous architect in the world, demanded an explanation. The steward went back to the wireless office and returned with the startling information that the *Titanic* was sinking and that the *Olympic* was speeding toward her to render assistance. Daniel Burnham sat back stunned, realizing then his friend, Frank Millet, may never reach New York to get the letter he had left waiting for him. Burnham now noticed the vibration of the ship, the up-and-down motion of the plunging bow. The *Olympic* was being driven past her rated speed, and now he knew why. The lives of the people on the *Titanic*, including his old friend, depended on how fast the *Olympic* could reach her sinking sister.

Chapter Twenty-Two

Birma

30 Minutes

On Sunday night, wireless operator Joseph Cannon was listening to the news from Cape Race to put into Monday's onboard newspaper on the Russian East Asiatic Company vessel *Birma*. Cannon was twenty-four and had just married before talking his position as junior wireless officer on the 4,859-ton *Birma*. He was not a Marconi man but worked for Marconi's rival, United Wireless Company of America. United Wireless used the De Forest wireless system, and Marconi employees were told to ignore all transmissions using this system.

Getting the news from Cape Race was the standard practice among all wireless men on the ships. The passengers loved the onboard newspapers, but it took a lot of time to translate the news streaming in Morse. Cannon was serving with Thomas G. Ward from Southampton, who was the more experienced operator and had turned into bed already. This was only the second trip on the *Birma* for Cannon, with the trip to New York marked by large gray swells, whitecaps, and waves that broke over the ship. The turnaround in New York had the *Birma* headed for Rotterdam and then to Libau, a Baltic seaport. The ship was mainly for cargo with a small number of passengers, and she left New York on Thursday at the same time the *Titanic* was leaving Ireland. Her speed topped out at 13 knots and she was heading toward the *Titanic*. The three thousand miles between the ships would be quickly reduced, and by Monday morning they would pass each other with the *Titanic* on a northerly course for

New York and *Birma* more to the south headed for Europe. On Sunday evening *Birma* was a thousand miles from New York and the signal from Cape Race was strong as Cannon transcribed the Morse code.

The static filled his headphones and then cleared. "CQD-SOS from MGY. We have struck an iceberg sinking fast come to our assistance. Position Lat 41 46 N, Long 50 14 W, MGY."[1] Cannon wrote down the message, recording the corrected position the *Titanic* was sending out. He didn't know the call letters of the ship but woke up Ward, who immediately sent back. "MGY, what is the matter with you? SBA."[2] Phillips tapped back. "Ok. We have struck iceberg and sinking, please tell captain to come—MGY."[3]

The two men then rushed to the bridge and gave the messages to Captain Stulping, who immediately deduced they were 106 nautical miles south southwest of the ship that Cannon could not find in his identification book because the *Titanic* was so new that she wasn't listed yet. Stulping gathered with his officers and plotted a new course on a heading of east northwest or 56.5 degrees by compass. He then ordered out fifteen more stokers to shovel coal and get the *Birma* rolling through the sea as fast as possible. Ward then tapped back. "We are 100 miles from you, steaming 14 knots be with you by 6:30 p.m. Our position Lat 40.48 N, Long 52.13 W SBA."[4] Jack Phillips knew 6:30 was way too late, of course, but he tapped back, "Ok OM-MGY."[5]

Operator Ward then tapped out a call to the *Frankfurt* asking who the MGY was that was sending out continuous SOS SOS CQD CQD. Ward waited and then came the reply. He wrote down the letters and turned to Cannon, shaking his head, looking down at the message he had just transcribed. Joseph Cannon read the words, not believing what he saw. "MGY is the new White Star Liner *Titanic*—Titanic-OM DFT."[6] The ship started to vibrate beneath the two men, and they understood then they were going to attempt to rescue the largest ship in the world.

CHAPTER TWENTY-THREE

The Sleeping Captain

20 Minutes

THE *CALIFORNIAN* TURNED SLOWLY UNDER THE BRIGHT PINWHEEL STARS. She was a sleeping ship on the frozen North Atlantic while just ten miles away panic and chaos was breaking loose. The engine was not running and so the creaks and groans of the steel ship could be heard as she slowly turned in the syrupy North Atlantic. Two hours before, donkey engine room man Ernest Gill had come up from the intense heat of the engine room just before midnight, when he "saw plainly over the rail on the starboard side the lights of a very large steamer about 10 miles away" and "her port side lights."[1]

He did not know it was the *Titanic*, but he readily made out that it was not a freighter or a small vessel. He then went back to his cabin and told his mate, William Thomas, "It was clear off to the starboard or I saw a big vessel going along at full speed."[2] Gill couldn't sleep and returned to the deck, where he "saw a white rocket about ten miles away on the starboard side and in seven or eight minutes saw distinctly a second rocket in the same place, saying to himself, 'That must be a vessel in distress.'"[3]

Gill would later give an affidavit to the committee in the United States investigating the tragedy where "he says the captain was appraised of these signals but made no effort to get up steam to go to the rescue. The *Californian* was drifting with the ice."[4] Gill became so indignant he tried to marshal a committee of men to protest Captain Lord's actions, but he could find no one to stand with him. So, while the *Titanic* sank,

Captain Stanley Lord lay on the settee in the chart room of the *Californian*. It was warm in the chart room and the slight lulling turning motion of the ship rocked him gently. He was stretched out in uniform with his arms crossed. The ship groaned occasionally in the ice-cold water, bumping growlers—large pieces of floating ice—tapping the steel hull like soft hammers.

Captain Lord wasn't going to cross the ice, even though a Senate Committee would find later that all might have been saved (aboard the *Titanic*) but for the "negligent indifference of the steamship *Californian* to the *Titanic*'s distress signals."[5] Captain Lord felt that if the ship had a problem, they would respond by Morse. No need to wake wireless operator Evans. And what if he did. What? He was going to risk the lives of his crew, his passengers, and himself to cross a field of icebergs at night? No. Whoever that fellow was, he would have to get himself out of his own predicament. He had obviously disregarded the ice warnings. And besides, maybe nothing was wrong. Maybe he was just firing off rockets. Lord felt the warmth of the electric heater in the chart room, somebody murmured something, then he was gone.

Up on the bridge, Apprentice Officer James Gibson kept his eyes on the ship that looked very queer. She seemed to be slowing down. He watched as another rocket spidered the sky. "Rockets or shells, throwing stars of any color or description, fired one at a time at short intervals"[6] were to be regarded as a signal of distress. Second Officer Herbert Stone was well aware of Article 31 from the International Rules of the Road. The ship he was watching was sending off rockets and appeared to be listing in the water. Her lights glittered in the glassy darkness. Stone sipped his tea and kept his eyes on the rockets that exploded above her masts. Three, four, five, six. The rockets kept coming. He estimated her to be less than ten miles away.

The *Californian* continued to slowly drift with a turn to starboard, her bow facing the distant ship with both her red and green running lights visible. To Second Officer Lightoller and Captain Smith on the bridge of the *Titanic*, the turning starboard and port lights meant she was coming. They had been firing rockets nonstop. They had been Morsing nonstop, but the ship remained where it was. Phillips was sending out CQD and

SOS nonstop. Nothing. The ship just sat there as if asleep. Even if they couldn't hear the boom of the rockets, surely they saw them. Smith had even muttered wanting a six-inch naval gun to "wake that fellow up."[7] The captain of the doomed ship didn't know how prescient he was.

Before he had fallen asleep, Captain Lord had told Stone to keep on Morsing, and he had but there was no response. Gibson had returned to the bridge, and the two men discussed the rockets again. But now something else was happening. The ship was slowly disappearing. Stone could no longer see her red sidelight. No, she was simply vanishing, as if someone was pulling a cloak over her. If she had been streaming away to the southwest, he would have seen her green running light. No, like the queer position of the ship itself, this had no precedent. The ship was vanishing before his eyes.

At 2 a.m., Stone sent Gibson down to the chart room to wake Captain Lord. "Tell him that the ship is disappearing in the southwest and that she had fired altogether eight rockets."[8] Gibson went below and knocked lightly on the chart room door. Lord was known to be brusque and mostly unapproachable to the men and the officers. Waking up a man like Captain Lord was a touchy affair. He knocked again lightly, heard a voice from within, then opened the door. He repeated Stone's message. Lord was quiet, then his voice came sleepily out of the dim room. "Were they all white rockets?" "Yes sir." "What time is it?" "2:05 by the wheelhouse clock sir."[9] Captain Lord then nodded to Gibson, clicked off his light, and went back to sleep. Gibson returned to the bridge and told Stone what the captain had asked him. He looked for the ship, now faintly visible. Gibson watched the ship, her lights fading like houses on a rural circuit slowly losing power. Then she just vanished.

RMS *Titanic* ready for launch, 1911.
WIKIMEDIA COMMONS

Harold Bride, the only surviving
wireless operator of the *Titanic*.
WIKIMEDIA COMMONS

Jack Phillips stayed at his post until the ship sank and then he drowned.
WIKIMEDIA COMMONS

Titanic's Marconi room, where Jack Phillips and Harold Bride transmitted the distress signals.
WIKIMEDIA COMMONS

The only known photo of the *Titanic*'s Marconi room with Harold Bride transmitting.

Captain Arthur Henry Rostron of the *Carpathia*, who rescued survivors of the *Titanic*.

The last collapsible lifeboat from the *Titanic* to be rescued by the *Carpathia*.
WIKIMEDIA COMMONS

Captain Edward J. Smith, who took the *Titanic* full speed into the ice field.
WIKIMEDIA COMMONS

Chapter Twenty-Four

That's the Way of It Now

10 Minutes

WHAT JACK PHILLIPS AND HAROLD BRIDE DIDN'T KNOW AS THEY tapped out the last wisps of electric current with the water rising all around them and the wireless room inverting like a rocket about to be launched was that the *Parisian* was only fifty miles away, but her wireless operator, Donald Sutherland, had gone to bed after spending all day trying to get assistance for the steamer *Deutschland*, which was disabled. Captain Haines had ordered Sutherland to bed at 10 p.m. The two wireless operators didn't know the closest ship was the *Californian* with its sleeping Captain Lord and two officers on the bridge watching the *Titanic* sink. They didn't know the *Mount Temple* was nosing around the far side of the ice field with crew and passengers watching the *Titanic* blast off her rockets while her captain refused to enter the ice.

They knew the *Olympic* was on her way. They knew the *Baltic*, the *Birma*, the *Virginian*, the *Frankfurt* were all steaming toward the *Titanic*. But the only ship that offered any real hope was the *Carpathia*, and now Phillips could not communicate with her. It was 2:15 a.m., and he was trying to coax out just a few more signals, which the *Virginian* picked up—and then his headphones went dead. The dynamos had been flooded, and the current on the ship was dying down like a candle fading to an orange ember. Bride had gone on one final inspection of the ship to assess their chances of finding a lifeboat. It wasn't good. He said, "I saw a collapsible boat near a funnel and went over to it. Twelve men were trying to boost it down to the boat deck. They were having an awful time. It was the last boat left. I looked at it longingly a few minutes, then I gave them a hand."[1] Bride then returned to the wireless room where Phillips was using his shoes to brace himself against the wireless table that was now itself sliding inch by inch forward. Worse than that, water had seeped into the wireless room and was now pooling around his ankles. The power had faded, with the lights turning sepia, and Phillips tried to adjust the spark to keep transmitting. *CQD MGY SOS SOS.*

The door opened and Captain Smith walked in. He saw Phillips hunched over his wireless key with Bride just behind him. He walked over to Phillips and looked at the operators. He spoke quietly. "Men, you have done your full duty. You can do no more. Abandon your cabin. Now it's every man for himself."[2] Phillips paused, glanced up, then went back to Morsing. The captain spoke again. "That's the way of it at this kind of time."[3] Then Captain Smith turned and left the wireless shack.

Phillips continued squeezing every bit of electricity out of his wireless set. He had been in communication with the *Olympic, Celtic, Cincinnati, Virginian,* and *Asian,* but they were all too far to do him any good. At 1:30 a.m., he had told the *Olympic,* "We are putting off passengers in small boats,"[4] and at 1:45, he relayed that "the engine room flooded."[5] Phillips's only hope was the *Carpathia* and maybe the *Mount Temple,* although he had not heard from her directly. He had sent the *Carpathia* a last message, "Engine room full up to boilers,"[6] at 1:46. He adjusted the spark again, hearing a terrific roar down in the bowels of the ship. The list was now down and to port as more water entered the wireless room. Phillips kept transmitting, looking for a miracle. His last signals might be the difference between life and death. His feet ached and were freezing in the water, but he kept tapping away, *CQD CQD SOS MGY.*

Captain Smith was now walking the boat deck and called to the men struggling with the collapsible. "You've done your duty men. Now it's every man for himself." To Fireman McGann, "Well boys, its every man for himself." Again, to oiler Alfred White, "Well boys, I guess it's every man for himself." To Steward Edward Brown, "Well boys, do your best for the women and children, and look out for yourselves."[7] Then he returned to the bridge and disappeared into history as some of the crew threw themselves into the water. A baker cannonballed into the water and years later would shudder when thinking of the freezing ocean that hit him like a million knives. Greaser Fred Scott belly flopped into the water, while some tried to dive close to lifeboats, hoping to be picked up. Stewards loitered around the rails of the boat decks in their white uniforms, speculating how long it would take for the *Titanic* to actually sink. First-class bellboys smoked while the gym instructor T. W. McCawley declared a life jacket would only slow him down.

Jack Thayer and his friend Long faced the water and knew it was time to go. Said Thayer,

We had no time to think, only to act. We shook hands, wished each other luck. I said, "Go ahead. I'll be right with you." I threw my overcoat off as he climbed over the rail, sliding down facing the ship. Ten seconds later I sat on the rail. I faced out, and with a push of my arms and hands, jumped into the water as far out from the ship as I could. When we jumped, we were only twelve or fifteen feet above the water. I never saw Long again. His body was later recovered. I am afraid that the few seconds elapsing between our going meant the difference between being sucked into the deck below, as I believe he was, or pushed out by the backwash. . . . The cold was terrific. The shock of the water took the breath out of my lungs. Down and down, I went spinning in all directions.[8]

Third-class passengers were now racing toward the rising stern like ants trying to escape a rising flood. "The crystal chandeliers of the *à la carte* restaurant hung at a crazy angle, but they still burned brightly, lighting the fawn panels of French walnut and the rose-colored carpet."[9] Just hours before, people had been in the Louis Quinze lounge and The Palm Court in formal attire having coffee and brandy after dinner, listening to classical music. Now it was the music of smashing dishes and tables sliding across rooms along with period furniture sodden with the weight of the freezing water. The entire ship was now a crazy house being tilted into the giant chute of the Atlantic Ocean. Thomas Andrews, the man who had designed the *Titanic*, remained in the first-class smoking room with his life belt on a chair, arms folded across his chest, staring at the painting *The Approach of the New World*. He simply could not believe the ship he had designed was now heading for the bottom of the ocean. He stared straight ahead not moving while the world tilted forward.

Lawrence Beesley sat in Boat 13 and watched the drama unfold:

The oarsmen lay on their oars and all in the lifeboat were motionless. . . . And then as we gazed awe struck, she tilted up slowly, revolving apparently about a center of gravity just astern of the midships, until she attained a vertically upright position, and there she remained—motionless! As she swung up, her lights, which had shone without a flicker all night, went out suddenly, came on again with a single flash, then went out altogether. And as they did so, there came a noise . . . partly a roar, partly a groan, partly a rattle, and partly a smash. . . . It went on successively for some seconds, possibly fifteen to twenty, as the heavy machinery dropped down to the bows of the ship.[10]

Second Officer Lightoller watched the people scramble up toward the stern away from the encroaching water. "It came home to me very clearly how fatal it would be to get amongst those hundreds and hundreds of people who would shortly be struggling for their lives in that deadly water."[11] Lightoller knew he was only delaying the inevitable and dove into the sea.

Striking the water was like a thousand knives being driven into one's body, and for a few moments, I completely lost grip of myself—and no wonder for I was perspiring freely, whilst the temperature of the water was 28 or 4 below freezing. Ahead of me the lookout cage on the foremast was visible just above the water. . . . I struck out blindly for this . . . till I got hold of myself and realized the futility of seeking safety on anything connected with the ship. I then turned to starboard, away from the ship altogether.[12]

Jack Phillips struggled to keep the set going, and at 2:10, the *Virginian* heard the last transmission of the RMS *Titanic* . . . two *V*'s. The

Caronia and the *Asian* heard distress signals but no response to their calls. Phillips was trying to coax the last bit of juice out of the set by using a lower power setting. Harold Bride began to think about what he would need to survive in the North Atlantic.

> *I thought it was about time to look about and see if there was any-thing detached that would float. I remembered that every member of the crew had a special life belt and ought to know where it was. I remembered mine was under my bunk. I went and got it. Then I thought how cold the water was. I remembered I had some boots, and I put these on, and an extra jacket, and I put that on. I saw Phillips standing out there still sending away.*[13]

Bride draped his life jacket over his friend's shoulders. The lights were now glowing a dim orange as Bride gathered up some papers. A woman was brought in who had fainted and placed in a chair until her husband came in and retrieved her. Bride returned to the bunk room behind the curtain to retrieve some of Phillips's money. "I looked out the door. I saw a stoker or somebody from below decks, leaning over Phillips from behind,"[14] he later told the *New York Times*.

> *He was too busy to notice what the man was doing. The man was slipping the belt off Phillips's back. He was a big man too. As you can see, I am very small. I don't know what it was I got hold of. I remem-bered in a flash the way Phillips had clung on—how I had to fix that life belt in place because he was too busy to do it. . . . I suddenly felt a passion not to let that man die a decent sailor's death. . . . I did my duty. I hope I finished him. . . . I don't know. We left him on the cabin floor of the wireless room, and he was not moving.*[15]

This story of the stoker seems incredible, but Bride later repeated the story for several newspapers, and it would seem to have the ring of truth. The younger operator then reached for the logbook, but Phillips shouted, "Let's clear out."[16] They dashed out of the wireless shack, and

Bride would later tell a reporter, "Phillips ran aft, and that was the last I ever saw of him alive,"[17] He wrote later:

> *I went to the place I had seen the collapsible boat on the boat deck and to my surprise I saw the boat and the men still trying to push it off. . . . I went up to them and was just lending a hand when a large wave came awash of the deck. The big wave carried the boat off. I had hold of an oarlock and went off with it.*[18]

Bride went into the bone-numbing cold and popped up in a cold, dark cavern. "I was in the boat, and the boat was upside down and I was under it. And I remember realizing I was wet through and that whatever happened I must not breathe for I was under water."[19]

Ten miles away, on the sleeping silent *Californian*, Third Officer Groves stood on the bridge and stared into the darkness where the lights of the ship firing distress rockets had been. The lights had flickered and then seemingly vanished. Now he saw only the coal-black night of a million stars. It was as if the ship had passed down into the ocean.

Chapter Twenty-Five

The *Times*

5 Minutes

A SLEEPING COPYBOY IN NEW YORK WAS AWOKEN AT 1:20 IN THE MORN-
ing on April 15, 1912. He was jolted awake when the dumbwaiter
slammed down the long shaft that connected the wireless room on the
roof with the editorial rooms of the *New York Times*. The copyboy took
out the paper with the transcribed message, read it, and ran for the editorial offices. The elevators seemed to take too long, so he went up the stairs,
reaching the managing editor's office and bursting in, his chest heaving.
Editor Carr Van Anda stood up and looked at the sweating, gasping boy
and read the message. He read it once, then he read it again. And again.
"From Cape Race, Newfoundland, Sunday night. April 14 AP at 10:25
o'clock (New York time) tonight the White Star Line steamship '*Titanic*'
called CQD to the Marconi station here and reported having struck an
iceberg. The steamer said that immediate assistance was needed."[1] By
now the copyboy had caught his breath, but Van Anda was now the one
breathing hard as he sat down in his chair and stared at the message.
This could be the biggest story of the century if true. He examined the
message. It came from the station in Newfoundland that had picked up
the distress call. He needed confirmation. He called the White Star Line
office in New York, then the *Times* reporters in Halifax and Montreal. He
began to piece together the story as it stood then:

A half hour before midnight (New York time) the Allen line had
received a transmission from their steamer, the Virginian, *which had*
picked up one of Titanic's *early distress calls and had altered course*

to rush to the stricken liner's aid. The White Star ships Olympic *and* Baltic *were also putting about, as were the* Birma, *the* Mount Temple, *and the* Carpathia. *Cape Race was monitoring the wireless transmissions between these ships as well. . . . Cape Race had heard nothing from the sinking liner since 12:27 a.m. (New York time), when a blurred and abruptly cut off SOS was heard.*[2]

The copyboy had left his office by the time Van Anda got off the phone. This was all he had to work with right now. The ship might be sinking, might have sunk; then again, it might not have sunk. But Van Anda had a nose for news, and the fact the *Titanic* had not been heard from and that one of the later transmissions was they were putting off women and children in lifeboats filled him with a deep sense of dread. This giant, unsinkable ship might well have gone down and taken an unimaginable number of people with her.

He would take a gamble. Van Anda immediately pushed the story covering the feud between Theodore Roosevelt and President Taft to the inside pages. The morning edition would lead with the *Titanic* sinking. He would hedge the truth and put into the paper what he knew but suggesting the worst. It was a brilliant move that would put the New York daily into a frontline position it would never lose. By taking a gamble, if his hunch was correct, he would scoop every other paper. Readers all over New York would wake up to a four-line headline.

New Liner Titanic *Hits an Iceberg*
Sinking by the Bow at Midnight
Women Put Off in Life Boats
Last Wireless at 12:27 a.m. Blurred[3]

Van Anda had one more trick up his sleeve. That was the morning edition, but with the city edition he would play his long hand and declare the *Titanic* had sunk. If he was right, then he would make history, if he was wrong, then he would be out of a job. The managing editor sat in his desk. He then turned around and stared out into the darkness. Somewhere out there, people were thrashing about in the icy Atlantic, fighting for their lives. If he was right, then many would die.

CHAPTER TWENTY-SIX

Only God Can Save You Now

1 Minute

LAWRENCE BEESLEY, THE SCRIBE IN THE LIFEBOAT, STARED IN STUPE-faction at the largest ship in the world as she began to rise up for her final plunge. With the discovery of the wreck in 1985, questions about Beesley's and others' accounts arose. Still, he was there. "And then, as we gazed awestruck, she tilted slowly up, revolving apparently about a center of gravity just astern of the midships, until she attained a vertically upright position and there she remained—motionless!"[1] And here is where modern technology belies Beesley's account. It is here that *Titanic* split in two from the pressure of the weight of the colossus standing up like a giant teeter-totter. "As she swung up, her lights, which had shone without a flicker all night, went out suddenly, came on again for a single flash, then went out altogether."[2] Beesley, for some reason did not see the ship split in two, but he describes a sound that could well be this moment:

> *There came a noise which many people, wrongly, I think, have described as an explosion; it has always seemed to me that it was nothing but the engines and machinery coming loose from the bolts and bearings, and falling through the compartments, smashing everything in their way. It was partly a roar, partly a rattle, and partly a smash, and it was not a sudden roar as an explosion would be; it went on successively for some seconds, possibly fifteen to twenty.[3]*

This, more than likely, was the moment the *Titanic* was shorn of her bow. Beesley even recognizes that accounts differ from his. "Several apparent authentic accounts have been given, in which definite stories of explosion have been related—in some cases even with wreckage blown up and the ship broken in two, but I think such accounts will not stand close analysis."[4] It is hard to say why Beesley did not see the ship separate. He might have looked away, and who could blame him, but he finished up by saying, "When the noise was over, the *Titanic* was still upright like a column; we could see her now only as the stern, and 150 feet of her stood outlined against the star-speckled sky, looming back in the darkness, and in this position, she continued for some minutes—I think as much as five minutes, but it may have been less."[5]

On board the ship, frantic people climbed up the mountain that the *Titanic* had become. Colonel Gracie headed for the stern but was blocked by a crowd of steerage passengers flooding up from below. Incredibly, the band reportedly still played and switched to the Episcopal hymn "Autumn," but this is hard to believe, as they literally would have fallen down the ship were they not hanging on; yet the mythology has the band playing on until the bitter end. A hysterical woman reportedly pleaded with anyone who would listen. "Oh, save me! Save me!" Peter Daly, the Lima representative of Haes and Sons, a London firm, shook his head, watching the water roil over the ship. "Good lady. Only God can save you now."[6]

Steward Brown fought against the water swirling around his knees as he pulled on Collapsible Boat A when the boat suddenly began to float off. He instantly jumped in and cut the stern lines and then another wave carried him away. The bow of the *Titanic* went straight down now as the ship slid forward into the sea and the stern rose like a giant seesaw. Beesley saw the final moments from his boat. "Then, first sinking back a little at the stern, I thought, she slid slowly forwards through the water and dived slantingly down, the sea closed over her and we had seen the last of the beautiful ship on which we had embarked four days before at Southampton."[7]

Marconi operator Harold Bride swam out from under the collapsible boat. "There were men all around me—hundreds of them. The sea was dotted with them, all depending on their life belts. I felt I simply had to get away from the ship."[8] Many people would report a disorientation from the freezing water. As Charles Lightoller put it, "I lost my grip on myself."[9] This was partially shock, and it lulled people for minutes as they watched, almost apart from themselves. Bride paused in the freezing water and stared at the ship.

She was a beautiful sight then. Smoke and sparks were rushing out of her funnel. There must have been an explosion, but we heard none. We only saw the big stream of sparks. The ship was gradually turning on her nose—just like a duck that goes down for a dive. I had only one thing on my mind—to get away from the suction. The band was still playing. . . . I swam with all my might. I suppose I was 150 feet away when the Titanic, *on her nose, with her after-quarter sticking straight up in the air, began to settle—slowly.*[10]

Second Officer Lightoller felt weighed down as he swam in the freezing water. He realized it was the "great Webley revolver still in my pocket"[11] that was weighing him down, and he sent it "on its downward journey."[12] A ventilator that was twelve feet across with a wire grating began sucking in water. The ventilator led to a shaft that went all the way to the bottom of the ship, and water rushed in the vortex and drew Lightoller in with it.

I suddenly found myself drawn, by the sudden rush of the surface water now pouring down this shaft and held flat and firmly up against this wire netting with the additional full and clear knowledge of what would happen if this light wire carried away. The pressure of the ship just glued me there whilst the ship sank slowly below the surface. Although I struggled and kicked for all I was worth, it was impossible to get away, for as fast as I pushed myself off, I was irresistibly dragged back, every instant expecting the wire to go and to find myself shot down into the bowels of the ship.[13]

To make matters worse, Lightoller was drowning and had just about given up hope when a "terrific blast of hot air came up the shaft and blew me right away from the air shaft and up to the surface."[14] Lightoller began to swim, but the water swirled again and he was sucked down and stuck to "one of the fiddley gratings."[15] He managed to get off the grating and wrote later, "I was rather losing interest in things, but I eventually came to the surface again."[16] He popped up next to the Engelhardt boat that Bride had ended up under. Lightoller was surrounded by drowning people now and grabbed onto a piece of rope trailing from the collapsible.

Jack Thayer had been trying to get away from the boat. "Swimming as hard as I could in the direction which I thought to be away from the ship, I finally came up with my lungs bursting but not having taken any water. The ship was in front of me, forty yards away."[17] Like Bride and Lightoller, Thayer seemed to be disorientated and paused to watch the *Titanic* sink. This would be understandable except he's floating in water that can kill a human being in fifteen minutes.

> *The ship seemed to be surrounded with a glare and stood out of the night as though she were on fire. I watched her. I don't know why I didn't keep swimming away. Fascinated, I seemed tied to the spot. . . . The water was over the base of the first funnel. The mass of people on board were surging back, always back toward the floating stern. The rumble and roar continued, with even louder distinct wrenching and tearing of boilers and engines from their beds.*[18]

Thayer then saw what was confirmed years later by the discovery of the *Titanic* on the floor of the Atlantic. Beesley would go out of his way to say the ship did not split in two, but Thayer describes what we know now to be true.

> *Suddenly the whole superstructure of the ship appeared to split, well forward to midship, and blow or buckle upwards. The second funnel, large enough for two automobiles to pass through abreast, seemed to be lifted off, emitting a cloud of sparks. It looked as if it would fall on top of me. It missed me by only twenty or thirty feet. The suction of it drew me down and down.*[19]

Thayer finally struggled back to the surface and put his hand up "against something smooth and firm with rounded shape."[20] He had his hand on the cork fender of one of the collapsible lifeboats with "four or five men clinging to her bottom."[21] It was the same one Lightoller was trailing with a piece of rope and Bride had found himself under. Thayer pulled himself up onto the boat but could not get his legs up and had to be helped up. He found himself sitting on the boat, exhausted, frozen, once again facing the *Titanic*.

Colonel Gracie thought the wave sweeping over the ship that had knocked Bride off was like the ones he used to surf as a boy in Newport. He rode the wave and rose up body surfing along until he grabbed onto an iron railing on the officers' quarters and then found himself temporarily beached on the roof under the second funnel. But the roof left him as the ship plunged deeper, and in the vortex of the rushing water into the steel cavern he became a spinning top caught in the whirlpool of a giant drain. He was still hanging onto the railing but finally let go and fought his way to the surface and managed to swim clear.

The wave took women, men, children, and babies. Chef John Collins had tried to help a woman from third class with her two children and had held the baby. He had guided her along with two other stewards to the port side where they heard a boat was and then went back to the starboard side. They paused there, thinking of heading for the stern when the wave swept over the woman, the two stewards, and Collins. He felt the baby pulled from his arms and never saw the woman, the children, or the stewards again.

Lightoller managed to get up on the overturned Engelhardt and watched the final moments of the ship he had served on. Then the real horror began. Lightoller heard a series of pistol shots. The steel anchor wires that held the forward funnel had snapped under the pressure from the ship's angle and now the funnel came down and slammed into the water on the starboard side in a cavalcade of sparks and twisting metal, flattening the swimmers alongside the ship and sending out a wave that carried Collapsible B thirty yards away from the ship. Greaser Walter Hurst, who was struggling in the water when the funnel fell, was blinded by soot, but he was one of the few to survive the crash of tons of steel.

Some lights still glowed dimly on the ship. Father Byles was at the very end of the stern with third-class passengers surrounding him. "Hail Mary, full of grace . . ." The stern rose higher yet, and the mass of humanity still on board, some 1,500 souls, pressed together toward the tip, clinging to the deckhouses, ventilators, cargo cranes, and hatch covers. "The Lord is with thee." The ship herself shrieked with agony, her hull subjected to stresses it was never designed to withstand. "Pray for us sinners."[22] And now people were hanging from anything they could grab onto as the giant building that had been a ship now became a skyscraper with her propellors emerging from the deep like glistening bronze beaters of some colossal mixing bowl. People fell like apples from a tree, sliding straight down into the green, strangely illuminated water. Here and there, people lost their grip, tossing, turning, flipping, going head down, sitting, some sliding along the deck, knocked senseless before the final plunge into the ice-cold water.

Third-class passenger Olaus Abelseth, his cousin, and his brother-in-law were back by the fourth funnel clinging to a rope like three puppets dangling over a precipice. Now the sea was rising up, and Abelseth held on until it was just below where they were hanging on the rope. He let go and immediately became entangled in ropes and believed he was doomed but suddenly kicked free and popped up outside of the ship.

From the lifeboats people watched in horror as the ship upended itself and then, incredibly, split in two, with the bow plunging away and the stern falling back into the water. The sound of the ship before and after it split apart had been described as the loudest thunder imaginable. Some called it a terrific roar. Walter Lord, the author of *A Night to Remember*, summed up the contents of the *Titanic* imploding upon itself as everything inside the ship broke loose. It is worth quoting, as it gives one the image of a luxury hotel suddenly upended.

> *There has never been a mixture like it, 29 boilers . . . the jeweled copy of the* Rubaiyat *. . . 800 cases of shelled walnuts . . . 15,000 bottles of ale and stout . . . huge anchor chains . . . 30 cases of golf clubs and tennis racquets for Spalding . . . Eleanor Widener's trousseau. tons of coal . . . Major Peuchen's tin box . . . 30,000 fresh eggs . . . doz-*

ens of potted palms . . . 5 grand pianos . . . a little mantel clock in B-38 . . . the massive silver duck press . . . the 50-phone switchboard . . . 2 reciprocating engines and the revolutionary low-pressure turbine . . . Billy Carter's new English automobile.[23]

It was a veritable slice of life at that time, a smorgasbord of classes and tastes and technology all crashing together in a terrific roar. To the 767 people in the boats, it was beautiful, horrible, terrible, ghastly, grotesque, fantastic, amazing, astounding, horrific. They watched as the people snaked up the ship toward the stern and then rode down with it when the ship cracked in two. And now they were watching again as the stern began to rise one more time. The lights flickered again but still amazingly burned on longer than anyone would have thought capable. And again, the *Titanic* rose, a glittering ship against the glittering stars dropping off people like ants falling from a giant tree. Some could not watch any more. Bruce Ismay turned away, and in Boat 1, C. E. Henry Stengel turned around, lamenting, "I cannot look any longer."[24] Elizabeth Eustis in Boat 4 put her hands over her face. It was just too much to watch. The lights flickered again, then flicked out, the ship now an oppressive black monolith against the pinwheel stars.

Thayer, lying atop an overturned collapsible, had a straight-on view of the inverted deck of the *Titanic*.

We could see groups of almost 1,500 people still aboard, clinging in clusters or bunches, like swarming bees, only to fall in masses, pairs or singly, as the great after part of the ship, 250 feet of it, rose into the sky, till it reached a 65- or 70-degree angle. Here it seemed to pause and just hung for what felt like minutes. . . . I looked upwards; we were right under the three enormous propellors. For an instant I thought they were sure to come down right on top of us.[25]

She began sliding then. A slab of steel slipping down into the freezing, syrupy water. The very best man had to offer at that point in time was simply swallowed up by a dark, flat ocean that left nothing more than a loud gulp to mark her passage. "She's gone. That's the last of her,"[26] someone remarked in Boat 13. And then after the loudest gulp in history, with only a strange, low-hanging fog to mark her passage, another sound began to grow. It would later be described as a collective moan, the sound of cicadas in summer, the roar of a crowd at a baseball field, a strange wail that was in all the same pitch. People who heard it said they could never forget the sound, and that it haunted their sleep, their dreams, and drove some to suicide. Whatever it sounded like, it was actually the collective scream of more than fifteen hundred people drowning in freezing water.

Twenty-two-year-old Margaret Schwarzenbach had been rowing around with her mother and father when the *Titanic* made its final plunge. She told her story fifty years later to a reporter:

> *Suddenly there was a terrible noise over the water as the great engines broke loose from their moorings and the Titanic tilted into the air and made a terrible plunge forward. It was during these moments that we heard the awful wailing. . . . It was the wailing of those left aboard in those last moments when the Titanic went down. It wasn't a screaming, not a shouting. It was an eerie, almost moaning noise. I recall some Shubert phrase . . . "a wailing over the water, a song of death."*[27]

CHAPTER TWENTY-SEVEN

His Power May Be Gone

0 Minutes

THE *CARPATHIA* HAD BEEN GOING FULL OUT, WITH THE ENTIRE SHIP vibrating like a creature that had been woken from a long sleep. Captain Rostron was risking his ship, his passengers, his crew, and his own life to come to the rescue of the *Titanic*. Rostron and his officers had organized the crew to be ready for the passengers plucked from the sea, and the lookouts were scanning the ocean for the icebergs that had ripped *Titanic* apart.

> *Down below in the boiler room, the stokers and firemen worked with renewed vigor, pressure gauges were pegged on every boiler. In the engine room, the big pistons of the* Carpathia's *reciprocating engines still pounded up and down in a blur, crankshaft throws spun in a flash of polished steel, link heads rocked to and fro, and eccentric rods flicked back and forth, steam belching from the cylinder heads with every stroke of the pistons. Every plate, frame, and rivet shook with the exertion as the* Carpathia *thundered on.*[1]

Marconi operator Cottam had heard nothing from the *Titanic* since the final message, "Come as quickly as possible old man; she is taking water, and it is up to the boilers."[2] After that, Cottam could only pick up static. On the German ship *Frankfurt*, Marconi operator Zippel had tried to contact the *Titanic*, but nothing came back. Captain Hattorff still pushed his ship to get to the *Titanic's* location as quickly as possible. At 1:47, the *Caronia* had detected weak signals from the *Titanic* but was unable to decipher the Morse code. Cape Race sent a message to the *Virginian* at 1:55: "We have not heard *Titanic* for about half an hour. His power may be gone."[3] At five minutes after 2 a.m., the Marconi operator on the *Virginian* heard static and then some faint signals from the *Titanic* but could not make out any of the letters. All that could be deduced was that the power of her set had been greatly reduced. The SS *Asian* thought they heard a faint SOS but then nothing else. At ten minutes past two o'clock the *Virginian* received the letter *V* from the *Titanic*. It came through twice, and the radio operator was sure it was the *Titanic* as he recognized the sparking background noise.

At 2:15, Captain Rostron stood on the bridge next to Second Officer James Bisset and told him to find two more lookout men and post the on the bow. His instructions were clear, look for icebergs but also look for the *Titanic*. The *Carpathia* was still going as fast as possible and had already cut around some growlers but had not slackened their speed. Rostron also ordered the lookout in the crow's nest to be replaced. He wanted fresh eyes to scour the ocean for ice. Purser Brown came to the bridge and reported all preparations had been completed and the ship was ready to receive any *Titanic* passengers they could rescue.

At 2:17 the *Virginian* again thought she heard the *Titanic*. The signal was *CQ CQ CQ CQ*. The operator waited for the following message, but none came. He wrote on his notepad, "Signals end very abruptly as if power suddenly switched off. His spark rather blurred or ragged."[4] He then sent a signal back suggesting the *Titanic* switch to the emergency set all the larger ships were equipped with. There was no response, and the *Virginian* asked the *Olympic*, "Have you heard anything about *Titanic*?"[5] The *Olympic* signaled back. "No. Keeping strict watch but hear nothing more from *Titanic*. No reply from him."[6]

Captain Moore on the *Mount Temple* had been nosing around the far side of the ice field that he had retreated from earlier. His passengers and crew still believed they had seen the *Titanic* and would later swear to many different newspapers they had seen the ship and the distress rockets. Captain Moore had roused his crew and set them to swinging out the lifeboats to be lowered and getting rigging lines ready. Moore didn't believe he should risk his own passengers lives in a quest to save another ship's passengers, but he also didn't want to be blamed for not going to the *Titanic's* aid. So, when at 2:30 a.m. the lookout reported some ice floating on the sea, he told the engineer in the engine room to stand by in case they had to suddenly reverse the engines and back away again. The ice was small blocks, so Moore continued eastward, but he was edgy. The iceberg that had sunk the *Titanic* was out there in the darkness, and he sent his fourth officer up to the bows with the directive for the lookout to keep a very sharp eye for any iceberg in the path of the ship. The fourth officer was to stay up there as well and look for ice. Captain Moore didn't like any of this. He was not Captain Rostron, who was zigzagging his way through icebergs at top speed. Like the still sleeping Stanley Lord, Captain Moore believed the ice posed imminent danger to his ship and believed more in Captain Smith's final order on the *Titanic* of every man for himself than the ethos of Rostron's heroic mission to reach the stricken ship. To the people in the freezing water fighting for their lives, they knew it was every man, woman, and child for themselves now. Their only hope was the diminutive captain on the *Carpathia* urging his ship on with every ounce of steam his boilers could muster. It was a matter of life and death and Captain Rostron believed every second mattered now.

Chapter Twenty-Eight

A Thin Smoky Vapor

10 Minutes after the Titanic *Sank*

THE RESCUE OF THE *TITANIC* PASSENGERS NOT ONLY INVOLVED THE SHIPS racing to the scene but also the passengers themselves. The *Titanic* had twenty lifeboats with a capacity of 1,178. Those twenty boats now ringed the spot where the ship had slipped beneath the glassy water. In place of the ship were now more than fifteen hundred people thrashing in the water, fighting off not only drowning, but also freezing to death. The lifeboats of the *Titanic* held only had 705 people. It was a colossal demonstration of incompetence on the part of the crew that the boats were not loaded to capacity, but they were not. In those twenty boats was room for 473 more people. All the boats had to do was come back.

Harold Bride could no longer feel his feet. His arms and legs felt like frozen, sodden logs. When he hit the water, he had involuntarily gasped, breathing in deeply and then felt a strange disorientation. It was the effect of 28-degree water assaulting his body. And he was a healthy young man. People succumbed to cardiac arrest from the shock of such icy water.

> *Survival time in water of twenty-eight degrees Fahrenheit is, at the very most, thirty minutes, with consciousness lost around fifteen and most deaths following within the next five. . . . Breath is commonly driven from the body by the shock at contact with water so cold, forcing an involuntary breath to be taken which can significantly increase the chances of drowning.*[1]

Bride heard a sound that would later be described as a despairing moan all in the same key. It was from those around him fighting for their lives. Where the *Titanic* had been was now a "thin, smoky vapor soiling the clear night. The glassy sea was littered with crates, deck chairs, planking, pilasters, and corklike rubbish that kept bobbing up to the surface from somewhere now far below."[2] Steward Brown saw a man tearing feverishly at his clothes, while Olaus Abelseth from steerage felt a man clamp his arm around his neck using him as a human life raft. Abelseth struggled free, then the man clamped him again, until he finally kicked him away.

Bride's overturned boat had become a target for many swimmers, along with Collapsible A. Abelseth managed to haul himself in along with others already half dead. The class system of the *Titanic* had finally broken down. In the collapsible was "tennis star R. Norris Williams Jr. lying beside his waterlogged fur coat . . . a couple of Swedes . . . Fireman John Thompson with badly burned hands . . . a First-Class passenger in BVDs . . . Steward Edward Brown . . .Third-Class Passenger Mrs. Rosa Abbott."[3] Collapsible A then drifted off, and the swimmers could no longer reach it. So, they turned their attention to Bride's overturned enclave Collapsible B.

They swarmed up over the white keel and tried to hoist themselves up. Some were simply too weak from the cold and slipped down into the ocean and disappeared forever. Bride could hear the clunking on the bottom of the boat above him and the clamoring of people calling for help. Second Officer Lightoller and Jack Thayer found themselves by the boat and were able to hoist themselves aboard. Walter Hurst had also managed to grab onto the keel and pull himself up. A man who looked like a great soggy bear swam up to the boat and with one motion flung himself on the keel. A. H. Barkworth, a Yorkshire Justice of the Peace, had worn his fur coat over his life vest, and this gave him some insulation against the agonizingly cold water. Colonel Gracie had clung to a wooden crank and then a plank and then swam toward the overturned collapsible. By the time he reached the boat a dozen men were either kneeling or lying on the boat. Gracie grabbed the arm of one man and hoisted himself aboard. Assistant cook John Collins surfaced from somewhere and climbed aboard as well.

Bride had watched his small cavern of air become smaller and smaller as the weight of the men pushed the overturned boat down. He knew it was now or never, but his body shook from uncontrollable spasms and his teeth chattered incessantly. Still, he could not stay under the boat any longer without drowning. He took a big breath and then dove down in the bone-chilling water and surfaced alongside the boat, surprised to see no less than thirty men now on Collapsible B. He managed to get onto the stern and almost passed out from the cold and exertion. Said Bride, "There was just room for me to roll on the edge. I lay there not caring what happened. Somebody sat on my legs. They were wedged in between the slats and were being wrenched. . . . It was a terrible sight all around—men swimming and sinking . . . others came near. Nobody gave them a hand."[4]

Steward Thomas Whitely was one of the last allowed aboard, and, even then, someone hit him with an oar, but he managed to get aboard. The upside-down collapsible carried more men than some of the lifeboats. They swung boards to get away from the other people and paddled as best they could. "Hold on to what you have old boy, one more would sink us all,"[5] the men on the boat shouted at the swimmers. Another man swam up and waved. "All right boys, good luck and God bless you!"[6] He then swam away.

Hurst turned and to his dying day swore it was the voice of Captain Edward Smith. He held out an oar to the man, but he turned away. They finally began to clear the swimmers, and a seaman on the keel asked the others if they should pray. They decided on the Lord's Prayer and began murmuring in the darkness. It did not drown out the sound that Fireman George Kemish would later describe as a "hundred thousand fans at a British football cup final."[7] Lightoller heard voices that would haunt his dreams forever. As the ship sank, he could hear "husbands and wives, brothers and sisters, parents and children crying out to one another, 'I love you.'"[8] When the *Titanic* had slipped beneath the waves, Third Officer Pittman, in Boat 5, quietly announced, "It's 2:20."[9] It had taken the RMS *Titanic* just 160 minutes to sink.

Lightoller tried to keep the overturned boat level as the water crept up higher and people dropped off into the sea. "Some quietly lost consciousness and slipped overboard, there being nothing on the smooth flat bottom of the boat to hold them. No one was in a condition to help, and the fact that a slight but distinct swell had started to roll up, rendered help from the still living an impossibility."[10] He watched the sea encroaching on the boat and took charge.

> *I realized that without concerted action, we were all going to be pitched headlong into the sea, and that would spell the finish for everyone. So, I made everyone face one way and then, as I felt the boat under our feet lurch to the sea, one way or the other. I corrected it by the order, "Lean to the right, "Stand upright," or "lean to the left."*[11]

Bride could no longer feel his frozen feet at all. He had told Lightoller about the *Carpathia* and said she should reach them by daylight. They just had to stay afloat until then. Thayer lay on top of the boat listening to the cries of the people in the water. It began with an "individual call for help, from here, from there, gradually swelling into a composite volume of one long continuous wailing chant. . . . It sounded like locusts on a midsummer night, in the woods in Pennsylvania. This terrible crying lasted for twenty or thirty minutes."[12]

Fifth Officer Lowe wanted to row back and rescue the people in the water. He had strung the lifeboats together. Boats 4, 12, 10, and D all were tied together about 150 yards away from the people in the water. Lowe decided it was best for one boat to go back to try and rescue the people. So, he began shifting people in the boats and divided up his fifty-five passengers and picked out the strongest oarsmen for Boat 14. It was 2:30 a.m. and Lowe knew time was of the essence and lost his temper more than once. "Jump, God damn you, jump,"[13] he shouted at Miss Daisy Minahan. Another woman in a shawl hesitated and Lowe pulled down the shawl and found a young man staring at him. He threw the man into Boat 10 and continued on. Still, it took time, and he wanted also to let the swimmers thin out so they didn't swamp the boat. It was more than forty minutes after the *Titanic* had sunk when Boat 14 moved in among the people in the water.

They picked up a Japanese passenger who was floating on a door, then first-class passenger W. F. Hoyt, then steward John Stewart. Lowe was frustrated. So many people he reached had already died, floating corpses who turned and bobbed in their life jackets. He called out and called out and would hear a voice, a shout, but then nothing. Four passengers were all he could find to rescue, and Hoyt died in his boat. Lowe had simply not understood how fast the freezing water sucked the life out of people.

Lawrence Beesley, in Boat 13, was shocked when he heard the first cries for help.

The cries of the drowning floating across the quiet sea filled us with stupefaction. We longed to return and rescue at least some of the drowning, but we knew it was impossible. The boat was filled to standing room, and to return would mean the swamping of us all, and so the captain stoker told his crew to row away from the cries. We tried to sing to keep all from thinking of them.[14]

The people in the other boats heard the cries as well. In Boat 5, Pittman turned the boat around and announced, "Now men we will pull toward the wreck."[15] But this was not to be. "Appeal to the officer not to go back,"[16] a woman begged Steward Henry Etches. "Why should we lose all our lives in a useless attempt to save others from the ship."[17] This would become the mantra of the people in the lifeboats. The Edwardian age of chivalry, valor, and honor had vanished over the hill. In its place were clawing people unwilling to help their fellow human beings freezing and drowning right in front of them.

Other women on Boat 5 protested, and Pittman had no choice but to tell his men not to row to the people in the water. "For the next hour, No. 5—40 people in a boat that held 65—heaved gently in the calm Atlantic swell, while its passengers listened to the swimmers three hundred yards away."[18] And then the situation reversed itself in Boat 6. Here the women wanted to return and help the drowning. Mrs. Lucien Smith, Mrs. Churchill Candee, and Mrs. J. J. Brown (the unsinkable Molly Brown) all asked Quartermaster Hichens to return and rescue the people in the water. Hichens would not. He said the swimmers would swamp the boat and they would all lose their lives. The women persisted as the wail began to fade, but Boat 6 never went back, with only twenty-eight people in a boat that held sixty-five.

Fireman Charles Hendrickson was behind the clarion call in Boat 1, saying, "It's up to us to go back and pick up anyone in the water."[19] Silence. No one said a word. He announced again that it was their responsibility to return and help the people in the water. Lookout George Symons who was in charge of the boat only looked down. Cosmo Duff-Gordon then cleared his throat and said that he didn't think it would be wise to return and that the boat might be swamped. Hendrickson realized he was outnumbered and fell silent. Boat 1 only had twelve people in it, with a capacity of forty. They rowed about in the dark with no direction.

Only Boat 14 returned to help the people in the water who had just an hour before been their fellow passengers. Otherwise, it did not matter to the people if they were first class, second class, third class, they were not returning. The cries for help were starting to die off very quickly. The twenty boats that had been launched floated around and rowed here and there but were distant from their fellow men, women, and children, who were dying all around. It might have been Guggenheim or Astor in the water, but it didn't matter. No one wanted to risk anything even for men who at the time of the sinking were in the top rung of society. No amount of money could help anyone in the water now. They were ignored along with immigrants who could not speak a word of English. In death, the class system of 1912 was finally abolished. There would be no inquiry either as to why the rescue of 407 people who could have been pulled into the boats did not occur. Almost a third of the people in the water might have been saved if it were not for the failings of human compassion—and once again, the very trait the *Titanic* mythology has promoted more than anything else, *courage*, was found sorely lacking when it was needed most. The lifeboats of the *Titanic* not returning to the drowning has always been the dark spirit of that night.

CHAPTER TWENTY-NINE

A Dead Silence

60 Minutes after the Titanic *Sank*

ON THE *CARPATHIA* IT WAS COLD. THE NOISE OF THE CREW AND THE preparations had woken up more than a few of the passengers. The engines seemed to be pounding so hard they felt like they might come up through the deck. But it was the cold that indicated something very strange was happening aboard the *Carpathia*. The ship was bound for Gibraltar, Genoa, Naples, Trieste, Fiume. In this pre-Florida era, her 175 mostly elderly passengers were headed for a sunny vacation along with 575 steerage passengers, mostly Italians and Slavs returning to their homeland.

Captain Rostron had been dodging ice for the last hour with his engines running faster than the boat engineers thought possible. Operator Cottam was still hunched over his wireless and was still in his shirtsleeves even though the heat to the ship had been turned off an hour before. He had been undressing when the first CQD came in and he never got around to putting his coat back on. The *Carpathia* was now entering the heart of the ice field, and Rostron was staring into the darkness, trying to see the next iceberg that might sink his ship.

At 2:40 a.m., Dr. McGee had joined him when he saw a glimmer of green light far off in the horizon. "There's his light!" Rostron exclaimed, "He must still be afloat!"[1] He knew ships had distinctive flares, and the White Star Line's was a dark green flare. Rostron assumed these large flares could not have been fired from a lifeboat and that must mean the

Titanic was still afloat. Second Officer Bisset then saw an iceberg lit by a reflected star and the ship had to veer around it, but she didn't slow down. Rostron was convinced the flare was from the *Titanic* and ordered colored rockets to be fired in intervals with Cunard Roman candles. The rockets would be fired in fifteen-minute intervals. He wanted anyone out there in the frozen darkness to know that help was on the way. Rostron began to pace the bridge, looking for another green flare to home in on.

The *Virginian* operator at 2:45 a.m. kept trying to get a signal from the *Titanic*. The *Baltic* picked up her signals and tried to raise the ship as well. Nothing. Operator Cottam tried signaling again, and sent "If you are there, we are firing rockets."[2] Cottam's message was picked up by other ships, but nothing came back from *Titanic*'s operator. At 3 a.m. the *Mount Temple*, which had been plying the ice field, saw the light of a ship directly in front of her. Captain Moore thought it might be the *Titanic* but then deduced it was the light of a sailing ship. He couldn't see the ship but assumed it was one of the fishing ships that took the plentiful cod off Newfoundland. The ship was going west, and Moore steered to starboard to avoid her when the light blinked out.

Captain Moore was squinting into the darkness, looking for the ship when a foghorn shook the windows of the bridge. Moore realized he was right on top of her. He had his engines slammed into reverse and turned the wheel hard to starboard. The *Mount Temple* shook and then halted. The foghorn blared again, and Moore saw the light again, but it was farther away. He turned away and headed back for the *Titanic*, once again encountering the mystery ship. The *Mount Temple* began to cautiously pick her way through the ice. What are we to take away from this?

The reports of mystery ships have dogged the *Titanic* mythology. There are two ways to go with this. The mystery ship scenario essentially lets everyone off the hook by saying the passengers, crew, Captain Smith, Second Officer Lightoller, people on board the *Mount Temple*, the *Californian*, people in lifeboats were all mistaking these large ships for rogue fishing vessels prowling the North Atlantic. This supposes people in the lifeboats were rowing for some mysterious trawler that wanted to remain hidden because she was breaking the law by fishing illegally. Ah, no.

The *Titanic* was a glittering floating palace of light at a time when illumination was just getting started on ships. The sheer size of the *Titanic* made her stand out, and that, along with her lights, meant mistaking her for a fishing vessel was like mistaking a Christmas tree for a night light. Add to this the fact that the night was unbelievably clear, the frigid air allowing light waves to go very far, and there is simply no way to think that a passenger or a crewman could mix up the behemoth that was *Titanic* and a ship that was barely a quarter its size. If anything, the idea that other ships were in the area breaks the lie that the *Titanic* was all alone and no ship could reach her. If we go with the mystery ship motif, then we have to give credence to the idea that the *Mount Temple* and the *Californian* and possibly others were all very close as the *Titanic* was in a major shipping lane. There might have been other ships, but there was no mistaking the *Titanic* for them. And no other ships were shooting up rockets. The people in the lifeboats were rowing for a stationary ship that seemed to be turning very slowly; the *Californian* was dead in the water doing a slow turn. This is the ship Captain Smith, Second Officer Lightoller, people in the boats, all saw just ten miles away or closer.

Operator Cottam on the *Carpathia* tried again at 3:10 a.m. to signal the *Titanic* and was picked up by the *Birma*. The Russian operator thought it was the *Titanic* and after speaking with his captain sent back, "Steaming full speed for you. Shall arrive 6 morning. Hope you are safe. We are only 50 miles now."[3] Cottam picked this up and assumed the *Titanic* was in contact with the Russian ship and tried frantically to make contact for the next half hour. Then at 3:15 a.m. the *Carpathia* signaled another steamer. She was still going full speed and weaving around the icebergs when a lookout shouted. Rostron saw a steamer bearing down on them with only masthead and side lights. There were no cabin lights and it seemed to be a small freighter. Rostron could not deviate from his course or his mission, as he knew he was close, and let the ship go on—another sighting of the mystery ship. Captain Moore, meanwhile, ever cautious, ordered his engines to go even slower as he hit more ice at 3:25 a.m. Moore was in the western edge of the ice field. He had backed away from the ice field once before and now was almost stopped again. The ship he had seen in the night was still ahead of him, steering to the south.

Captain Rostron on the *Carpathia* was not only worried about his ship but others as well. He was dodging around one iceberg after another and he thought of the other ships coming full speed that might blunder into the ice field. He was very concerned about the *Titanic*'s sister ship *Olympic*, which he knew was coming at 23 knots or even faster. He had Cottam send a message to Captain Haddock and gave him a direct ice warning. "South point pack ice 44.46 North. Do not attempt to go north until 49.30 West. Many bergs large and small, amongst pack; also, for many miles eastward."[4]

The airwaves were once again flooded with messages going back and forth among the ships and being picked up by the shore stations. The night sky of the North Atlantic was a range of electromagnetic waves bouncing from one ship to another and sometimes the messages were truncated and the meaning changed entirely. The information that the Cunard liner *Caronia* was in contact with the *Titanic* took hold, and she was besieged with requests for information.

Meanwhile, on the *Californian*, Chief Officer George F. Stewart relieved Second Officer Stone on the bridge. It was 4 a.m. and Stone described the night to him, the strange ship that appeared to be listing, the rockets, and then the ship's disappearance, and finally telling Captain Lord of all of these events. Stewart was listening and raised up his binoculars and turned southward seeing a "four-masted steamer with one funnel and a lot of light amidships."[5] She hove into view and Stewart asked Stone if this was the ship he had seen the night before. He said it was not, and "he was sure it was not the same one that had fired the first eight rockets."[6] Then Stone went below and left Stewart on the bridge. Stewart raised his binoculars again and stared at the ship that appeared to have stopped. He had an uneasy feeling that something bad had happened.

On the *Carpathia*, Captain Rostron had reached the location of the *Titanic*. At 3:50 a.m., he rang down "Stand By" to the engine room. At 4:00 a.m., he sent down an "All Stop." His ship had reached 41.46 N, 50.14 W. This was the position that Phillips had sent to the *Carpathia* operator Cottam and everyone else on the high seas. This was where the *Titanic* should be. But there was nothing. Nothing but the absurdly calm water. And a dead silence.

CHAPTER THIRTY

The Stillest Night Possible

100 Minutes after the Titanic *Sank*

THE *TITANIC* WAS GONE. AND SO WERE THE PEOPLE DROWNING IN THE
freezing water. They were now frozen corpses kept afloat by their white
life jackets. They floated off slowly with the North Atlantic current,
many never to be seen again. A strange lull fell over the area after the
ship slipped beneath the waves, leaving only a low-hanging fog. Now a
stupefying silence replaced the last cries and moans for help, for God, for
mother or father. Second-class passenger Lawrence Beesley felt the loss
of the *Titanic* as if an old friend had passed away.

> *In place of the ship on which all our interest had been concentrated for
> so long and towards which we looked most of the time because it was
> the only object on the sea which was a fixed point to us—in place of the*
> Titanic *we had the level sea now stretching in an unbroken expanse
> to the horizon, with no indication on the surface that the waves have
> just closed over the most wonderful vessel built by man's hand.*[1]

There was simply nothing except the detritus of chairs, boxes, crates,
and other objects—amazingly few things considering the enormity of the
ship that had gone to the bottom of the ocean.

Jack Thayer, on the overturned boat, later wrote about the boats not
returning to help the drowning.

The most heartrending part of the whole tragedy was the failure right after the Titanic *sank, of those boats which were only partially loaded, to pick up the poor souls in the water. . . . The partially filled lifeboats standing by, only a few hundred yards away, never came back. Why on earth they did not come back is a mystery. . . . They were only 400 to 500 yards away, listening to the cries, and still they did not come back. How could any human being fail to heed those cries?*[2]

But Thayer had his own problems. The boat was slowly sinking.

During this time, more and more were trying to get aboard the bottom of our overturned boat. We helped them on until we were packed like sardines. . . . There were finally twenty-eight of us. We were very low in the water. The water had roughened up slightly and was occasionally washing over us.[3]

People were literally lying on top of each other on the small boat. Thayer wrote,

We were standing, kneeling, lying, in all conceivable positions, in order to get a small hold on the half-inch overlap of the boat's planking, which was the only means of keeping ourselves from sliding off the slippery surface into that icy water. I was kneeling. A man was kneeling on my legs with his hands on my shoulders, and in turn somebody was on him. Once we obtained our original position we could not move. The assistant wireless man, Harold Bride, was lying across, in front of me, with his legs in the water, and his feet jammed against the cork fender, which was about two feet under water.[4]

Harold Bride kept everyone's spirits up by telling them which ships had answered his CQD and that "the *Carpathia* is coming up as fast as she can. I gave her our position. There is no mistake."[5] They had broken into song to try to drown out the screams of the dying. Officer Lightoller later wrote that the "heartrending, never-to-be-forgotten sounds . . . seemed to go through you like a knife."[6] Thayer later said the singing was a ruse: "It would help us keep our bearings, but we all knew it was only a kindly ruse to try and drown out that awful moaning cry. . . . God Grant that I never hear such a song again."[7]

But by 2:30, silence fell. The last person had succumbed to hypothermia in the freezing water and now the people in the boats were left to themselves and the guilt of their own inaction in helping their fellow human beings. Lady Rothes had no such guilt in Boat 8. She had wanted to go back and retrieve the people struggling in the water nearest their boat. She knew that the "cold of the water was so awful that very few could bear it alive for more than a few minutes."[8] She told the people in her boat they should row back and rescue some of the people but was met with a stony silence. Only the sailor in charge, Thomas Jones, and Gladys Cherry and an American woman supported her in going back. Cherry later said, "The other women and the two stewards would have killed us rather than go back."[9] Sailor Jones gave in but covered himself by saying, "Ladies, if any of us are saved, remember I wanted to go back. I would rather drown with them than leave them."[10] Cherry would later write about not going back to save people as the "dreadful regret I shall always have."[11]

The awful silence that followed presented a stunning indictment of all those who had not gone back and rescued the people in the water. These survivors, whom history would treat kindly, were presently being judged by the harshest judge of all, their own conscience. For in the following silence the floating life-jacketed bodies were now a testimony to their inaction. Dorothy Gibson, a silent movie star aboard the *Titanic*, remembered that after the *Titanic* had sunk and the last person had been silenced by the cold, "for the next two hours, the passengers in our boat just sat in the darkness and tried to keep warm. No one said a word. There was nothing to say and nothing we could do."[12] Cherry would forever be haunted by that night on the Atlantic and described how it was the "stillest night possible, not a ripple on the water and the stars wonderful; that icy air and the stars I never want to see or feel again."[13]

Elizabeth Weed Shutes in Boat 3 watched shooting stars streak across the sky and wondered how the rockets from the *Titanic* could have been visible to anyone. The stars were just fast streaks of light and then there were only the brilliant stars that reflected off the still, glass-smooth water. Fourth Officer Joseph Boxhall added his own star to the sky by firing off a green flare. He began to fire them off at timed intervals, and some people in the farther boats thought they might be rescue ships. Sound carried in such immeasurable silence. The water offered no barriers, and voices along with the splashes of oars, the clunking of oarlocks, people calling to one another crossed the black space between the boats. Some boats tied up for safety and comfort. Boats 6 and 16, along with 5 and 7, tied up together while other boats drifted farther apart. Twenty boats were in the middle of the North Atlantic floating about in a radius of six miles, some rowing, some just floating along. The placidity of the water and the strange tranquility of the moment after so much death and destruction lulled the passengers in the boats. One stoker blurted out, "It reminds me of a blooming picnic."[14]

There were children. Babies. Old people. Young people. Many newly widowed women who were realizing their husbands were gone. Beesley tried to wrap a baby in a blanket, while Edith Russell tried to keep a baby amused with a toy pig. Three-year-old Louis Navratil was fed cookies by Hugh Woolner, while Mrs. John (Jack) Astor gave a shawl to a Swedish woman for her daughter. It would have had the air of a picnic if it were not for the bone-chilling cold. Beesley remarked on the absolute darkness surrounding them. "None of the other three boats near us had a light, and we missed lights badly; we could not see each other in the darkness, we could not signal to ships which might be rushing up full speed from any quarter to the *Titanic*'s rescue."[15]

In Boat 5 Third Officer Pittman wrapped a sail around a Mrs. Crosby because she was shivering so hard. The unsinkable Molly Brown draped her sable stole around a stoker whose teeth chattered like small hammers. A man in white pajamas in Boat 16 looked so cold and pale it seemed someone had frosted his entire body. In Boat 14 Mrs. Charlotte Collyer fell over from the cold and her hair caught in an oarlock leaving large tufts of hair. A sailor gave up his socks to Mrs. Washington Dodge, while in Boat 13 Fireman Beauchamp refused to take a coat offered to him by a woman and gave it to a young Irish girl.

The women in Boat 4 took the oars, Mrs. John B. Thayer rowing for five hours with water over her ankles, and in Boat 6 Molly Brown organized the women with two to each oar, one holding it in place and the other pulling hard. They rowed three or four miles toward the elusive light on the horizon that never seemed to get any closer. In Boat 2 Mrs. Walter Douglas took the tiller from Boxhall and helped fire off the green flares. An electric cane with a light at the end was used by Mrs. J. Stuart White in Boat 8 to signal other boats. The Countess of Rothes in Boat 8 had her maid row as she manned the tiller.

The realization that husbands, fathers, brothers, and sons were gone came in waves. Mrs. Charles Hays called out to every boat, "Charles Hays are you there?"[16] One woman screamed for her husband over and over in Boat 8 until the Countess of Rothes turned the tiller over to someone else and sat next to the woman and calmed her down by talking to her all night. After the fear of imminent death had passed, human frailty and pettiness moved in. In Boat 3 women bickered while their husbands sat silent. Smoking was an irritant, for in 1912 it was viewed by the upper classes at least as something not done in the company of ladies. Elizabeth Shutes asked two sailors to quit smoking; they ignored her. Mrs. Stuart White would even cite the smoking in the American investigation that followed, "As we cut loose from the ship these stewards took out cigarettes and lighted them. On an occasion like that!"[17] Sir Cosmo Duff-Gordon gave Fireman Hendrickson a cigar in Boat 1 and no one said a word because Lady Duff-Gordon was busy vomiting from being seasick, and Mrs. Francatelli, the only other woman on the boat, was employed by Lady Duff-Gordon. So Boat 1 was deemed a smoking boat.

But an incident in Boat 1 would forever tar Duff-Gordon with an accusation of paying people off to not go back and rescue those in the water. Boat 1 was already suspect, with only twelve people in a boat designed to hold sixty-five. And they were all first class. Fireman Pusey was mulling over the loss of his kit (toiletries, miscellaneous personal items). Duff-Gordon heard Pusey grumbling to his wife about the loss. Pusey abruptly turned to Duff-Gordon. "I suppose you have lost everything?" "Of course." "But you can get more?" "Yes." "Well, we have lost our kit, and the company won't give us any more. And what's more our pay stops tonight."[18]

Duff-Gordon stared at the working-class man, detecting his seething resentment of his position and his wife, who earlier had been consoling Mrs. Francatelli on the loss of her nightgown, which to the fireman seemed ridiculous. "Very well, I will give each of you a fiver to start a new kit."[19] He gave the men the money, and this became emblematic of class privilege, where Duff-Gordon controlled the boat through money while those in the water drowned. He would be pilloried in the press for all that was wrong about the rescue of those on the *Titanic*, where the first class was clearly the winner. It was implied that Duff-Gordon bribed the seamen not to return to the *Titanic* to rescue people in the water, but this has never been proven.

During the five hours after the sinking, the boats wandered about the dark ocean with no idea when and if help would arrive. The Countess of Rothes later wrote, "One began to feel rather hopeless,"[20] adding that the silence moved in and left them with the "ghostliness of our feelings."[21] Those who didn't want to deal with survivor's guilt of the remorse over not going back had one escape—drink. In Boat 4 a sailor became intoxicated from a bottle of brandy he had in his pocket and became obnoxious until the quartermaster threw him into the bottom of the boat.

In Boat 6 a mutiny of sorts occurred while the *Titanic* had yet to sink beneath the water. Major Arthur Godfrey Peuchen was in the boat and, as was his habit, he started to give orders. Quartermaster Hichens was in charge and at the tiller and Peuchen was pulling on an oar. The major asked Hichens to join him in rowing and let a lady steer the boat. Hichens essentially told Major Peuchen to shut up, declaring he was in charge and that he should be quiet and row. Only Peuchen and lookout Fleet were rowing, attempting to escape the sinking ship, while Molly Brown was getting the other women to help with the rowing. Hichens stayed on the tiller and yelled at the passengers to row harder or they might be sucked down. The women rowed toward the elusive light on the horizon until it was clear they were getting no closer. It was then that Hichens declared they were doomed, having no water, food, or charts and being hundreds of miles from the nearest landfall.

The women then took control. Major Peuchen had given up, saying they were lost. He stopped rowing, but the women told him to get back to it. "Mrs. Candee grimly showed him the North Star. Mrs. Brown told him to shut up and row. Mrs. Meyer jeered at his courage."[22] Molly Brown took charge and told a stoker who had transferred from Boat 16 to start rowing. Hichens tried to stop her, and she told him if he got in her way, she would throw him overboard. Hichens retreated to a blanket, hurling insults, but Boat 6 had a new captain, and her name was Molly Brown.

CHAPTER THIRTY-ONE

A Beacon of Human Failing

100 Minutes after the Titanic Sank

CHIEF OFFICER STEWART ON THE CALIFORNIAN WAS STILL ON THE bridge at 4 a.m. pondering the night's events. The ship that had appeared, the rockets, the second ship he had seen. Something had happened. Rockets were a distress signal first and last. Second Officer Stone said he thought the ship was less than ten miles away. Why Captain Lord had not gotten up to investigate he had no idea. But Stewart could not get out of his head that something ominous had occurred, and when he woke Captain Lord at his accustomed hour, he began to recite the events Stone had relayed to him until Lord cut him off. "Yes, I know," he snapped. "He's been telling me."[1]

Lord put on his great coat, scarf, and hat and headed for the bridge. Once there, he began to plot a route though the ice field. Stewart was frustrated and asked Lord if he was going to try to find out what happened to the ship firing off the rockets. Captain Lord raised his binoculars and stared at the four-masted steamer to the southwest, murmuring, "No, she looks alright. She's not making any signals now."[2]

Stewart did not mention that this was not the ship from the night before Stone had identified. Lord did not like to be contradicted, and in Stewart's mind he believed it was easier for him to try to find out about the ship than to get his captain to do something. They stood on the bridge for the next hour when Stewart excused himself and went down to the wireless room and woke up Cyril Evans. He shook the wireless officer awake who peered up at him in the dim light.

"Sparks, there's a ship been firing rockets in the night. Will you see if you can find out what is wrong. What is the matter?"[3] Evans stared at him for a moment then got up quickly, pulled on his pants and a pair of slippers, sliding down behind the wireless set. He wound the wire detector and turned on the brush motor and put on the headphones. He heard static, buzzing, then tapped out CQ-any station, come back. Stewart stood over him and waited. Evans sat for a moment when his headset blazed to life with a series of quick buzzing dots and dashes. He transcribed the letters. It was from the *Frankfurt*. "Do you know the *Titanic* has sunk during the night, collided with an iceberg."[4]

Evans looked up Stewart, who was staring at the pad of paper he had written the message on. He tapped back and asked what her position was. "41.46 N, 50.14 W." The *Virginian* then cut in. "Do you know the *Titanic* has sunk?" Evans quickly replied, "Yes, the *Frankfurt* has told me."[5] Evans tapped back again. "Please send me official message regarding *Titanic*, giving position."[6] Evans wanted an official confirmation for Captain Lord. The position came back matching the position the *Frankfurt* had just given. He transcribed the signal from the *Virginian* and Stewart scooped it up and ran for the bridge with Evans close behind him, shouting that a ship had just sunk. Stewart reached Captain Lord and held the message out. Lord read the message then looked at Stewart and frowned. "No, no this can't be right," he said, handing the message back to Evans. "You must get me a better position than this."[7]

What Captain Lord was saying was that the RMS *Titanic* could not have sunk directly in front of his boat. There must be some mistake. There had to be a mistake because if there was not a mistake, then the *Titanic* sunk last night and was visible from his ship and probably was just under ten miles away. The rockets that Stone and Apprentice Officer Gibson had observed were from the *Titanic*. The ship that looked so strange as if she were heeling over was *Titanic*. The ship that had simply disappeared was the *Titanic*. The ship that sunk while Captain Lord slept and had not woken Evans to find out why a ship was firing distress rockets was the *Titanic*. In that moment, even though Captain Lord wished it away, saying you must bring me a better position, the die was cast, and the knowledge that the *Californian* could have saved the fifteen hundred people freezing to death in the water just hours before was on the bridge now too, along with the three men.

Evans left and went back to the wireless office and soon returned with another confirmation. 41.46 N, 50.14 W was the only position for the *Titanic*. He returned to the bridge and Captain Lord accepted it this time and started to work out a course for the *Californian* to reach the position. He had forgotten about the ship that was south to southeast. If he asked Evans to get back on the wireless and find out about the ship, he had seen the answer would have come back immediately. It was the *Carpathia* that had come from fifty miles away at full speed. Captain Rostron had come and endangered himself, his ship, his passengers, dodging icebergs, driving his ship to the limit on a knight's errand to save whomever he could. Captain Rostron was the antithesis of the man slowly plodding a course through the ice two hours after fifteen hundred people had drowned. Captain Lord had slept not ten miles away while his officers watched the RMS *Titanic* sink. The elusive light those in the lifeboats saw as a light of hope was really a beacon of human failing.

Chapter Thirty-Two

Let Us All Pray to God

105 Minutes after the Titanic *Sank*

THE MEN ON COLLAPSIBLE BOAT B WERE RUNNING OUT OF TIME. THEY were not idly floating around, smoking cigars, bickering with other passengers, getting drunk. The thirty men who were balanced on the bottom of the boat were balancing the collapsible to save their lives. The air was seeping out from under the hull and the boat was slowly going down into the water. The sea slopped over the keel as the men stood trying to balance on the equivalent of a very unstable teeter-totter. One false move and they would all be pitched into the sea.

Second Officer Lightoller was doing what he did best and took control. Lightoller had the thirty men stand in a double column, facing the bow. Then, as the boat lurched with the sea, he shouted, "Lean to the right. . . . Stand upright. . . . Lean to the left."[1] He was trying to counteract the slow rolling swells that passed over the boat every few minutes. Colonel Gracie was one of the men on the boat and his teeth chattered incessantly. Someone passed him a flask to warm him up, and he passed it to Walter Hurst, who swigged it, thinking it was brandy, but he nearly choked on the peppermint schnapps.

The men talked of getting rescued. Lightoller asked Harold Bride by the stern what ships he had contacted. Bride shouted back, the *Baltic*, *Olympic*, the *Carpathia*, *Frankfurt*, *Mount Temple*. . . . He told Lightoller the *Carpathia* was the closest and had responded, coming full speed. Lightoller calculated she should arrive in the morning. The men started to watch the horizon, even as they heard a sudden splash which meant someone had given up and could no longer stand the cold, the shivering, the pain of frozen feet and hands. The men simply dropped into the Atlantic like petals from a wilting flower. The green flares sent up by Fourth Officer Boxhall cheered them as they watched the horizon for any sign of a ship.

As dawn came so did a breeze that froze the men even more. Bride could no longer feel his feet and was so tired he was past caring and wondered many times what had happened to Jack Phillips. With the breeze, the sea went from the millpond the *Titanic* had sunk in to a choppy white-foaming expanse. Icy waves splashed over their feet and soaked the men up to their knees. No, this was not sustainable. Still the men fell away. They would do a half twist and then there was a desultory splash. No one said anything. They couldn't afford to. Their strength was too precious in the battle to stay afloat and live. The men just died around them without comment without expression. The black sea around them was silent and empty.

And now comes the story of the chief baker, Charles Joughin, who was still in the water and still alive. His ordeal began around 12:30 when he started drinking whiskey in his cabin. He and the other bakers had distributed bread on the boat deck and helped load women from the promenade deck. Then Joughin went back to his cabin and drank some more. Now the *Titanic* was listing heavily, and all the boats were gone. Joughin began throwing chairs from the enclosed promenade. He threw almost fifty chairs overboard. He then joined all the other passengers at the top of the stern as the *Titanic* made her final plunge. Like taking an elevator all the way down, he rode the big ship and then just stepped off into the water and didn't even get his hair wet. He swam off keeping his head above water, insulated from the cold by the whiskey. At 4 a.m. he saw some distant wreckage that turned out to be Collapsible B, but it was too full of people for him to climb on. The drunk baker hung on to an old friend, John Maynard, an entrée chef, who extended his hand and allowed him to tread water. He would survive.

Jack Thayer now noticed the air was gradually leaking from under the boat, "lowering us further and further into the water."[2] They had to do something. Lightoller blew his whistle to signal other boats to come take some of the people off the overturned collapsible. Two boats responded and slowly worked its way over and took the men off the overturned collapsible. Thayer wrote of the agonizing wait as the boat sank lower. "One by one, those on top of the freezing group stood up, until all of us who could stand were on our feet, with the exception of poor Bride, who could not bear his weight on his but could only pull his legs slightly out of the water."[3] Thayer could now see the lifeboat approaching. "We could see that so few men were in them that some of the oars were being pulled by women. . . . The first took half of us. My mother was in this boat. . . . The other boat aboard took the rest of us."[4]

But now the new boat was overloaded with seventy-five people, and the sea was beginning to kick up. At 3:30 a.m. there was a flash and a boom. A Miss Norton cried out, "There's a flash of lightning."[5] But Hichens, the quartermaster who had lost command of his boat, muttered, "It's a falling star."[6] A stoker lying unconscious in Boat 13 jumped up, shouting, "That was a cannon."[7]

Lightoller saw a light on the horizon, followed by another. He wiped his eyes, making sure the light was not a star. If the light was a star, then the stars were increasing. One light appeared, then another, and another until there were rows of lights stacked on top of each other. Charles Lightoller knew then it was a steamer making for them, and in Boat 9 deckhand Paddy McGough called out, "Let us all pray to God for there is a ship on the horizon and it's making for us!"[8] A newspaper was lit in Boat 3 and in Boat 13 letters were twisted together into a torch. Boxhall shot off a last green flare as Mrs. J. Stuart White swung her electric cane. Salvation was coming for the survivors of the *Titanic* as the dawn brimmed pink and strangely beautiful. Jack Thayer looked up in the overloaded lifeboat. "Sure enough, shortly before four o'clock, we saw the masthead light of the *Carpathia* come over the horizon and creep toward us. We gave a thankful cheer. . . . Even through my numbness I began to realize that I was saved—that I would live."[9]

CHAPTER THIRTY-THREE

She's at the Bottom of the Ocean

110 Minutes after the Titanic *Sank*

ON LAWRENCE BEESLEY'S BOAT THEY HAD A FALSE ALARM WHEN THE Northern Lights were mistaken for a ship. The first sign of the *Carpathia* was when "someone in the bow called our attention to a faraway gleam in the southeast," said Beesley. "We all turned quickly to look, and there it was certainly, streaming from behind the horizon like a distant flash of a warship's searchlight; then a faint boom like guns afar off and the light died away suddenly."[1] Everyone in the boat turned and "waited in the absolute silence in the quiet night. And then creeping over the edge of the sea where the flash had been, we saw a single light and presently a second below it."[2]

Captain Rostron on the bridge of the *Carpathia* was staring into the total darkness of the North Atlantic. He had reached the exact position radioed to Evans and yet there was nothing here. No *Titanic*. No debris. No bodies. Just ice and darkness. The lights of the bridge reflected up under his eyes as he raised his binoculars. Could they all have perished? Could the *Titanic* have sunk so fast and so horrifically that no one was able to survive. But that made no sense. Then a green spit of light. The flare arced up like a slash of green phosphorus and then was swallowed by the darkness again. Rostron guessed that was no ship's flare but only the little sparks of light he had seen before and had hoped was the *Titanic* but then had realized it was either a star or the refraction of light between the horizon and water. The clear pane of glass that was the sea had made it a night of optical illusion. But this flare was directly in front of him. He trailed it down to the water and in the pale green light like the flash of a dull negative he saw a lifeboat. Another green flare and now Rostron could clearly see the lifeboat and behind it he thought there might be others. The people in Beesley's boat frantically signaled the approaching ship as well. "We searched for paper, rags, anything that would burn. A hasty paper torch was twisted out of letters found in some one's pocket, lighted, and held aloft by the stoker standing on the tiller platform."[3]

Captain Rostron rang down, "Slow Ahead," and swung the boat to port when the lookout, Bisset, saw a large iceberg dead ahead. He immediately swung to starboard and went around the iceberg. The lifeboat was on the windward side, and a choppy breeze bobbed the lifeboat up and down. Rostron grabbed a megaphone and went outside to the starboard bridge wing. The first voice of the moribund *Titanic* called out in the darkness, "We have only one seaman in the boat and can't work it very well." "All right,"[4] Rostron shouted back, nudging the liner slowly closer to the boat he could barely see. He then ordered Bisset to go to the starboard gangway with two quartermasters and direct the lifeboat as it came along the ship. "Fend her off so she doesn't bump, and be careful that she doesn't capsize."[5]

"Stop your engines!"[6]

It was the strong voice of Fourth Officer Joseph Boxhall in Boat 2, which was drifting closer to the gangway. A woman's voice followed. Mrs. Walter Douglas of Minneapolis could not contain herself and said the words she had been wanting to say all night. She wanted to declare to the world what had happened, and she wanted to make sure the world understood the facts right now. She had been sitting next to Boxhall and stood up, shouting, "The *Titanic* has gone down with everyone on board!"[7] This shrill woman's voice in the frozen darkness was the first official confirmation from an eyewitness that the great liner had in fact sunk. Boxhall immediately told her to shut up, but she had said it and who knows how many people aboard the *Carpathia* had actually heard her. Now passengers and crew could see the boat clearly. Mrs. Ogden could see the White Star emblem on the boats and the strange way the life preservers made everyone looked as if they had on a white tuxedo. The only sound was the clunk of the oars and the sailors calling out. Mrs.

Crain saw the "pale, strained faces looking up at the decks."[8] What had they seen? A baby cried in the boat, sending chills through two women. These were real people who had survived something unimaginable, and many others had been lost, including children.

Lines were dropped and the boat tied off, and then a rope ladder rolled out from the gangway with a lifeline that Boxhall could tie under the arms of the passengers before they began their climb up the side of the ship. The first passenger rescued from the RMS *Titanic* was Elizabeth Allen, who was lifted onto the *Carpathia* by Purser Brown when she neared the gangway. It was 4:10 a.m. Brown asked the inevitable question. What happened to the *Titanic*? Allen gasped that the ship had sunk. Others followed, many too numb with cold to climb up the ladder. Boxhall was the last out of the boat and first officer of the *Titanic* to be rescued. Captain Rostron sent word that he needed to see him immediately on the bridge. Boxhall made his way to the bridge and stood before the captain, shivering uncontrollably. Rostron knew the answer before he asked the question, but he needed official confirmation.

"The *Titanic* has gone down?"

"Yes. . . ." Here Boxhall's voice broke with emotion. "She went down at about 2:30."

Rostron paused.

"Were many people left on board when she sank?"

"Hundreds and hundreds! Perhaps a thousand! Perhaps more!"[9] Boxhall went on, his voice breaking as grief began to get the better of him.

My God sir, they've gone down with her. They couldn't live in this cold water. We had room for dozens of people in my boat, but it was dark after the ship took the plunge. We didn't pick up any swimmers. I fired flares. . . . I think that the people were drawn down by the suction. The other boats are somewhere near.[10]

Captain Rostron nodded, and Boxhall left for the first-class dining room, which had been converted into an aid station. It was the knowledge the captain of the *Carpathia* didn't want. He had risked his ship, his crew, his passengers, his own life in a herculean attempt to reach the

Titanic before she sank, but in this he had failed. The *Titanic* had sunk, and Rostron knew there were nowhere near enough lifeboats for the twenty-two hundred souls aboard. Many had died in the Atlantic. This he knew. Now all he could do was rescue the lucky few who had made it into the lifeboats before they froze to death or succumbed to injures from the sinking.

Dawn was creeping over the North Atlantic now, and from the bridge Captain Rostron could see the small specks of gray that were the boats spread out over four or five miles. The captain could also now see something else. "In the growing light, maybe five miles off to the west, stretching from the northern horizon to the western, was a vast, unbroken sheet of ice, studded here and there with towering bergs, some as much as two hundred feet high."[11] To Rostron, it was like seeing a thief or murderer revealed. Among these pink edifices of morning light turning blue and gray was the iceberg that had sent the most advanced and largest ocean liner in the world to the bottom of the sea.

Charles Hurd had woken on the *Carpathia* and missed all the drama the night before but now could see from his porthole an ocean of menacing icebergs all around the ship. He dressed quickly and found a stewardess, demanding an explanation as to why they were stopped in the ocean and surrounded by icebergs. The stewardess pointed to some women making their way to the dining room and wiped her eyes. "They are from the *Titanic*," she told Hurd. "She's at the bottom of the ocean."[12]

The dawn came quickly, and the sky was clear, with the *Carpathia* emerging as a savior from the darkness to the people in the boats who had passed the time rowing, sitting, thinking, crying, freezing, dying. Beesley experienced the relief felt by many in the lifeboats as the steamer came to a stop.

> *We waited, and she slowly swung around and revealed herself to us as a large steamer with all her portholes alight. I think the way those lights came slowly into view was one of the most wonderful things we shall ever see. It meant deliverance at once. . . . I think everyone's eyes filled with tears, men as well as women.*[13]

Some of the people shouted joyfully, while others had organized cheers. Beesley's Boat 13 sang "Pull for the Shore, Sailor" as they rowed toward Rostron's ship. The ordeal was far from over. With the dawn, Captain Rostron could now see Boats 4, 10, and 12 with Collapsible D tied alongside. Men were soaked and in various stages of hypothermia. It was now or never. They could not row toward the *Carpathia* and might well sink; while salvation was at hand, the boats were now overloaded. Rostron directed the seaman toward the *Carpathia*.

Fifth Officer Lowe seemed to have been the only one who used the sail on Boat 14. All the boats had sails, but no one else had thought to hoist this obvious source of propulsion. At the Senate inquiry, Lowe would later explain why he was the only one to use the sail: "A sailor is not necessarily a boatman, neither is a boatman a sailor."[14] Lowe was a sailor, and he had his lifeboat clipping along at four knots toward the *Carpathia*. Not only was Lowe the only one who went back and actually rescued four of the swimmers, he now was picking up other boats in distress. He saw Collapsible D low in the water and tacked toward her when someone called out that they couldn't take any more people. They thought Lowe was looking to unload some of his passengers, but he told them to tie up to 14 and he would tow them toward *Carpathia*. They did, and Lowe caught the wind again, pulling the collapsible behind him. But he wasn't done yet. He saw Collapsible A low in the water about a mile away. When he reached the boat, he found out that half of the thirty people aboard had frozen to death and fallen overboard. Only thirteen survivors remained. He took the people off and put them into Boat 14 and took the collapsible in tow and then sailed once again for *Carpathia*.

"Oh Muddie, look at the beautiful North Pole with no Santa Claus on it,"[15] little Douglas Spedden said to his mother, Mrs. Frederic Spedden, as Boat 3 weaved through the ice. The ice sparkled all around the boats in a strange fantasy land of frozen beauty. The North Atlantic was waking up, and Lawrence Beesley would later write about the beauty of the morning:

> *And then, as if to make everything complete for our happiness, came the dawn. First a beautiful quiet shimmer away in the east, then a*

soft golden glow that crept up stealthily from behind the skyline . . .
then the sky turned faintly pink . . . and with the dawn came a faint
breeze from the west.[16]

In Boat 7, lookout Hogg encouraged the women by declaring, "It's all right ladies, do not grieve. We are picked up."[17] But the women, many who had lost their husbands, remained silent.

Now the twenty boats were converging on *Carpathia* in the morning sun.

It was 4:45 when Boat 13 tied up at the portside gangway, a half
hour after that when Boat 7 pulled alongside. At 6 a.m., survivors
from Boat 3 began to climb aboard Carpathia. *Some used the rope*
ladders, children were hoisted up in mail sacks, and some of the
women not strong enough to negotiate the rope ladders were lifted
aboard in slings.[18]

Elizabeth Shutes was hoisted up and heard a sailor shout, "Careful fellows, she's a lightweight!"[19]

Then, of course, there were moments of first-class privilege. Henry Sleeper Harper entered the gangway with his wife, his Pekinese, and his dragoman Hammad Hassab, and bumped into Louis Ogden, who had been sure the rescue of the *Titanic* was a cover story. Harper walked over to his old friend Ogden and said, "Louis, how do you keep yourself so young?"[20] Ogden was speechless.

The *Carpathia* crew stayed focused on the rescue of the *Titanic* passengers while Second Officer Bisset saw a four-masted steamer with a pale red funnel eight miles to the north-northwest of the *Carpathia*. First Officer Dean, who had been looking for more lifeboats, also saw the ship. Fourth Officer John Geoffrey Barnish and Third Officer Eric Rees also saw her. She seemed to be just starting and heading west into the ice field. A ship that would have obviously been visible to the doomed souls on the *Titanic*. It was the *Californian*.

CHAPTER THIRTY-FOUR

I Saw Rockets on My Watch

160 Minutes after the Titanic *Sank*

CAPTAIN LORD SAW THE *CARPATHIA* AS WELL, ALTHOUGH HE HAD NO interest in establishing her identity. Lord was busy trying to plot the safest way to the last position of the *Titanic* given to him by Evans. He rang for a "slow ahead." No need to get bothered by the fact that the men on his ship had just witnessed the sinking of the largest ocean liner in history and had stood by while their captain slept. No. Slow ahead. Don't want to put anyone in danger now that everyone had already drowned. Of course, Captain Lord didn't know this, he only knew that he was in a really bad spot. So, in another strange move, Lord decided to steam west through the ice field and out to the clear water on the other side. Instead of heading directly for the *Titanic*'s last position where there just might be people in boats who were freezing to death, he decided to take a circuitous route around the known position. Why? Maybe it was safer. "Steaming into the icefield then actually took the *Californian* further away from the scene of the *Titanic*'s disaster rather than bringing her closer."[1] Captain Stanley Lord was clearly not the man to have nearby when one was sinking.

Even with the updated information of the *Titanic*'s position from her discovery in 1985, he still was moving away from her position. Lord obviously regarded the scene as a death trap, and he slowly picked his way through the ice that had drifted in during the night. He did not want to leave anything to chance; those *Titanic* passengers could wait. He meandered along at four knots. The *Mount Temple* saw the *Californian* at 6 a.m. moving through the ice field six miles north of his ship. Lord finally found clear water at 6:30 a.m. and increased his speed to fourteen knots. Wireless operator Evans would unwittingly prove how close the *Californian* was to *Titanic* when she sunk, when at 6:10, the *Virginian* contacted Evans and asked for information on the *Titanic*.

Evans tapped back that the *Carpathia* was in sight and picking up people from the lifeboats. "This meant that even before 6:30 a.m., the *Californian* was already within seven miles—visual distance—of the wreck site, though she was by this time on the other side of the ice field because of Captain Lord's inexplicable maneuvering."[2] The *Californian*, even though she went to the far side of the ice field, could still see the *Carpathia*, and Lord himself must have conveyed this to Evans. The same way Second Officer Stone saw the *Titanic* sinking the night before, Lord could now see the rescue operation taking place. Either way, the *Californian* had been very close to the *Titanic* and still was very close. Captain Lord knew he was late to the party but made a point of turning out his whole crew.

He posted additional lookouts and lifeboats were swung out. Third Officer Groves stopped by Stone's cabin and had begun connecting the dots. He asked Stone if it was true about the *Titanic*. Stone nodded slowly. "Yes, old chap, I saw rockets on my watch."[3] Groves then proceeded to the bridge, where Captain Lord had steamed through eight miles of open water and reached the position of the *Titanic*.

He saw nothing. No lifeboats. No bodies. No *Titanic*. He saw the *Mount Temple*, and six miles to the east was the *Carpathia*. Captain Moore on the *Mount Temple* had also been making his way through the ice. He too had followed the position and 41.44 N, 50.14 W, but he and his lookouts saw nothing except for the ice to the north, south, and east. There was no sign of the *Titanic*. There was no debris either. Moore decided that the position had been wrong and started east, picking a way though the ice field. He and Captain Lord both headed for *Carpathia*. Lord once again took a circuitous route to reach the rescue ship. "Ice phobic" might aptly describe Lord at this point. He wound around the ice field and approached from the southwest.

Wireless operator Evans aboard the *Californian* had been tapping out messages while other ships on their way to rescue the *Titanic* looked for information. The result was so much cross chatter that a Marconi inspector, Gilbert Balfour on the *Baltic*, reprimanded Evans for his incessant chatter and silenced him for the rest of the day. Finally, at 8:30 a.m., the *Californian* reached the *Carpathia*. Captain Stanley Lord had finally come to the rescue of the *Titanic*. The *Carpathia* had raced across fifty-eight miles in under four hours and arrived one hour and forty minutes after the *Titanic* sank. The *California* had been ten miles away and didn't move at all, then, at a blazing four knots, arrived six hours and ten minutes after *Titanic* had sunk. The mysterious *Mount Temple* arrived too, the ship that had pulled away from the *Titanic* when she needed her most. It wasn't exactly the cavalry.

CHAPTER THIRTY-FIVE

We Believe That the Boat Is Unsinkable

Four Hours after the Titanic *Sank*

DAVID SARNOFF, IN HIS WIRELESS ROOM IN MANHATTAN, HAD BEEN trying all night to get more information. The *Parisian* had sent a message to the Cape Race Station, which had relayed it onward: "I have no survivors of the *Titanic* on board and no official information as to the fate of the sinking vessel."[1] The electromagnetic waves shooting around the ocean bounced to the land stations, and the world found out in bursts of Morse code. Wireless operators became the ears for the world on the night of April 14, 1912. In Pittsburgh, the night had been rainy and foggy. At the corner of 6200 Pennsylvania Avenue in the Vilsack building A. J. Edgecomb and W. D. Pyle also picked up the message revealing to the world the *Parisian* had not managed to reach the *Titanic* in time. The two men sat in the dim light of the incandescent bulb hanging from the ceiling, looking for any bursts of Morse code that might tell her story. Since the news of her sinking had broken in the early edition of the *New York Times*, the entire world was now yearning to find out what exactly happened and, more importantly, who had been saved.

Edgecomb and Pyle stayed at their posts throughout the night with the cold April rain pattering on the roof while they adjusted their tuner and snatched "aerograms" from the night. Their headsets caught fire in the early hours of April 15 when a burst of Morse lit up their set. They intercepted the message from Newfoundland that the *Titanic* had sunk at 2:20 Monday morning and no ship had been able to reach her. Edgecomb wrote quickly "that the women and children who had been lowered in lifeboats were rescued by the *Carpathia* three hours after the *Titanic* sunk."[2]

The men were in shock. All over America people were waking up to the devastating news that the RMS *Titanic* had sunk. And now they knew there had been a great loss of life. On *Titanic*'s sister ship the *Olympic*, people were waking up to find their ship thundering though the Atlantic with the lifeboats uncovered and ready for use. The journalist Ella Wheeler Wilcox was eating breakfast with her husband when they found out why the *Olympic* was humming with speed. "At the breakfast table our steward told us that news had been received that the *Titanic* had struck an iceberg but was saved with all on board. . . . He said, however, he feared more serious news might come later."[3]

Novelist R. H. Benson also got the news onboard and wrote in his diary later:

> *On coming down from mass this morning I heard a sentence from a lift boy that made me wonder. . . . I asked what the matter was and was told that the* Titanic *had communicated with us that she was in a sinking condition, that we were moving full speed towards her and should probably arrive at about 3:30.*[4]

Benson then plied information from other officials and learned that the *Olympic* had received the news sometime just after midnight and that *Titanic* was already sinking and her passengers were being put off into boats. He learned that two other ships were standing by, and one of them was the *Baltic*. Another passenger wrote a letter home when he learned the news:

> *The sea has quieted down some, and the sun is shining today. . . .* Titanic, *a sister ship, is in distress off the coast of Newfoundland. We have altered our course since three o'clock, and we are racing to her at full speed. Are making 25 knots an hour and expect to reach her about three o'clock this afternoon.*[5]

Another passenger on the *Olympic*, Robert Wilcox, told his son of a curious dream he had after waking from an afternoon nap. "I had dreamed of *Titanic*," he related. "I thought I saw it sailing over a smooth sea and then suddenly run up the sheer side of an enormous iceberg and turn a somersault and sink into the sea."[6] These passengers were among the few on the *Olympic* who actually knew what was happening.

Most passengers knew something had happened to the *Titanic* but that she was safe and the *Olympic* was rushing to help her. They saw that the stewardesses had prepared beds for the *Titanic* passengers and that water was to be conserved for the survivors. Actress Madame Simone thought the news was almost celebratory and envisioned the *Titanic* passengers coming aboard as a festive occasion. "At this news there was increased gayety on the *Olympic*, and cabins were prepared for the expected guests as if it were a festive occasion. In fact, much happy enthusiasm was shown in anticipation of these casual meetings in midocean."[7]

Vice President Philip Franklin of the International Mercantile Marine did not feel festive. He did not have much more information than the passengers on the *Olympic*. After the call from the reporter in the middle of the night he had summoned everyone he could to the New York office on Broadway. He was handed a memo when he arrived.

Titanic. *Received from Associated Press from Cape Race 3:05 a.m. Monday, April 15, 10:25 p.m. EST* Titanic *called CQD, reported having struck an iceberg and required immediate assistance. Half an hour afterwards, reported that they were sinking by the head. Women were being put off in boats and weather calm and clear. Gave position as 41.46 north, 50.14 west. Stop this station. Notified Allan liner* Virginian, *who immediately advised he was proceeding toward scene of disaster. Stop.* Virginian *at midnight stated was about 170 miles distant from* Titanic *and expected there about 10 AM.* Olympic *at 4.24 p.m. GMT in latitude 40.32 north, longitude 61.18 west, was in direct communication with* Titanic *and is now making all haste toward her.* Baltic *at 1:15 a.m. EST reported himself as about 200 miles east of* Titanic *and was also making toward her. Last signals from* Titanic *were heard by* Virginian *at 12:25 a.m. EST. He reported them blurred and ending abruptly.*[8]

Franklin knew a tsunami of reporters was headed his way, and soon the relatives of *Titanic* passengers would produce a second wave. From the Guggenheims to the poorest immigrants, they would all be clamoring for information, and at this point he really had none to give. He knew the *Titanic* was 1,080 miles from New York, and the *Olympic*, he and his staff estimated, was 360 miles east of her sister ship. Franklin wired the *Olympic* at 6 a.m., "Keep us fully posted regarding *Titanic*."[9] Captain Haddock had his wireless operator send an immediate reply: "Since midnight, when her position was 41.46 north, 50.14 west, have been unable to communicate. We are now 310 miles from her, 9 a.m. under full power. Will inform you at once if hear anything."[10]

Franklin essentially had no real news to share with relatives or the media at this point. New Yorkers awoke to see three different stories plastered across the front pages of newspapers.

The New Titanic *Hit by Iceberg, Appeals for Aid, Says the Wireless Report* (New York Herald)[11]

All Saved from Titanic *after Collision* (Evening Sun)[12]

New Liner Titanic *Hits an Iceberg, Sinking by the Bow at Midnight, Women Put Off in Life Boats, Last Wireless at 12:27 AM Blurred* (New York Times)[13]

Captain Haddock had the same problem and needed more information. Wireless operator Harold Cottam tried to contact Cape Race but could not understand her reply. At 7:50, Cottam tapped out a question for the *Asian*: "Can you give me any information *Titanic*, and if any ships standing by her?"[14] Cottam then asked the *Scandinavian* if she had any information regarding the *Titanic*. A half hour later, at 8:30 a.m., the *Asian* replied:

CAPTAIN Olympic*:* Asian *heard* Titanic *signaling Cape Race on and off from 8 to 10 p.m. local time, Sunday. Messages too faint* *to read. Finished calling SOS midnight. Position given as latitude 41.46 longitude 50.14. No further information.* Asian *then 300 miles west of* Titanic *and towing oil tank to Halifax.*[15]

At 8 a.m., there were crowds gathering outside White Star's office on Broadway with questions shouted at Franklin and his associates by reporters and relatives. He knew nothing and could only wonder along with the reporters and loved ones of the *Titanic* passengers what had really happened out there on the North Atlantic in the middle of the night. But he had to say something.

One hour later, the vice president of IMM explained to reporters that even if the *Titanic* had struck an iceberg, her watertight compartments would keep her from sinking. "We place absolute confidence in the *Titanic*," he said. "We believe that the boat is unsinkable. . . . The boat is unsinkable, and nothing but inconvenience will be suffered by the passengers."[16]

Olympic wireless operator Cottam, in the middle of the ocean, had become a lifeline of information for Franklin and the world, and at 9:25 a.m. New York time, he received an update from the SS *Parisian*:

I sent traffic to the Titanic *at 8:30 last night and heard him send traffic to Cape Race just before I went to bed. I turned in at 11:15 ship's time. The* Californian *was about fifty miles astern of us. I heard the following this morning at six o'clock. According to information received the* Carpathia *has picked up about twenty boats with passengers. The* Baltic *is returning to give assistance. As regards* Titanic *have heard nothing. Don't know if she has sunk.*[17]

Captain Haddock had Cottam relay the information to Franklin's office in New York. "*Parisian* reports *Carpathia* in attendance and picked up 20 boats of passengers, and *Baltic* returning to give assistance. Position not given."[18] Franklin took this information and sent back. "HADDOCK, *Olympic*: April 15, 1912. Thanks, your message. We have received nothing from *Titanic* but rumored here that she is proceeding slowly

Halifax, but we cannot confirm this. We expect *Virginian* alongside *Titanic*, try and communicate her."[19]

Franklin would not give up on the notion the *Titanic* was unsinkable and ordered a special train to go to Halifax to pick up *Titanic's* passengers. Relatives rushed to board the "*Titanic* Special." Furthermore,

> by 10 a.m., Franklin was fielding calls from Mrs. John [Jack] Astor, J. P. Morgan Jr., and Mrs. Benjamin Guggenheim. He told them all the same thing, that the ship was unsinkable, and all the passengers would probably be taken to Halifax. It was an amazing fantasy, but Franklin was not alone in his belief that the Titanic simply could not founder. The truth was that the world could not yet get its head around what had happened to the Titanic. In a classroom in England a headmistress said, "Stand up," any child who has a relative on the Titanic.[20]

And the entire class stood up. And one little mite said, "Oh there's no need to worry Miss, the *Olympic* is rushing to her aid."[21]

On the *Olympic*, Captain Haddock posted a bulletin in the first-class smoking room and first-class reading room. Just as the passengers were sitting down to have their lunch, they could be reassured that all was still right with the world. The bulletin read, "New York reports all passengers safe."[22] But the passengers in the freezing boats around the *Carpathia* knew only too well that the unsinkable ship had sunk into the ocean in 160 minutes. And she had taken many souls with her.

CHAPTER THIRTY-SIX

Pink Icebergs All Around

Five Hours after the Titanic *Sank*

FIVE-YEAR-OLD WASHINGTON DODGE LIKED THE RIDE. HE WAS GOING up through the freezing morning air and could see pink icebergs all around him. It had been an amazing night. He had seen the big ship he was on break apart and go down into the ocean. He had seen lots of people in the water after the ship sank. Then he had floated around in a lifeboat for many hours while the adults rowed. Then he saw fireworks again just like the rockets that had exploded over the *Titanic*. And he had just been plucked out of his boat and was in a cloth mail sack being hauled up high above the new ship that had just pulled up, and now he was going back down and landing gently on the deck. Dodge Junior stood up and a steward walked up and offered him coffee. He demurred and asked if he might have some hot chocolate. The steward nodded and went off to find the young master some hot chocolate.

Then a man with pajamas peeking out over his leather slippers in a long black overcoat appeared on deck. He had made it up the ladder but just barely. He looked to be first class with his baronial mustache and tousled brown hair. He looked a little like someone who had been pulled away from a dinner party and now he stumbled around with vacant eyes. He leaned against a bulkhead, his entire body trembling. Dr. McGee approached the director of the White Star Line. "Will you not go into the saloon and get some soup or something to drink?"[1]

"No. I really don't want anything at all."

"Do go and get something."

"If you will leave me alone, I'll be much happier here," Bruce Ismay blurted out, then changed his mind. "If you can get me in some room where I can be quiet, I wish you would."

"Please," the doctor softly persisted, "go to the saloon and get something hot."

"I would rather not."[2]

Dr. McGee gave in and took Ismay to his own room, where he never left. He didn't eat and didn't see anyone and remained sedated. Ismay would leave the White Star Line within a year and live out the rest of his years a recluse on a vast estate on the coast of Ireland. The cabin of Dr. McGee was the first bulwark against the world, but he still had a role to play in the rescue of the *Titanic*. The passengers now were flooding onto the *Carpathia*'s decks. Fifth Officer Lowe's little flotilla sailed up and all the passengers clambered aboard the *Carpathia*. Third-class passenger Olaus Abelseth hit the deck around seven and was given a hot blanket and then some brandy and hot coffee in the dining saloon. Lowe, before he left the boat, stowed his mast and sail, ever the efficient sailor. There was a happy reunion for Dr. Dodge when he found his wife and son after being separated on the *Titanic* when he insisted they get into a lifeboat. He had been lucky enough to be ordered into one himself to man an oar.

As the sun rose higher, the sea around the *Titanic* survivors revealed itself. Lawrence Beesley thought he saw two more ships, but "in a few minutes more the light shone on them, and they stood revealed as huge icebergs, peaked in a way that really suggested a ship." He continued, "When the sun rose higher it turned them pink and sinister as they looked towering like rugged white peaks of rocks out of the sea . . . the sun came above the horizon, they sparkled and glittered in its rays."[3]

The survivors on the promenade deck immediately became passengers staring down as other survivors disembarked from the lifeboats, looking for loved ones. A Mr. Carter from Boat 4 saw his family but not his son. "Where's my son, where's my son?" he asked, panicked.[4] A girl with a large hat turned her face up and exclaimed, "Here I am father." The story was that John Jacob Astor had put the hat on the ten-year-old after being denied entry to the boat, saying, "Now he's a girl, and he can go."[5]

Carpathia's passengers were now coming out of their staterooms and lining the rails, but some still did not realize they had stopped for a rescue operation in the middle of the North Atlantic. As Beesley wrote later, he imagined the scene the passengers saw: "Out ahead and on all sides little torches glittered faintly for a few moments then guttered out—and shouts and cheers floated across the sea."[6] A knock woke Mr. and Mrs. Charles Marshall in their stateroom. "What is it?" called Mr. Marshall. "Your niece wants to see you sir,"[7] came the answer. Marshall paused. His three nieces were making the maiden voyage on the *Titanic* and had sent him a wireless telegram the night before. It was impossible for any of them to be aboard the *Carpathia*. The steward explained to the dumbfounded Marshalls that his nieces had just been hoisted onto the *Carpathia* from one of the lifeboats of the *Titanic*. A family reunion was later held with Mrs. E. D. Appleton and all three nieces who eventually arrived.

Beesley, in Boat 13, had to row around several icebergs to reach the *Carpathia*.

We rowed up to her about 4:30 and, sheltering on the port side from the swell, held on by two ropes at the stern and bow. Women went up the side first, climbing rope ladders with a noose around their shoulders to help their ascent, men passengers scrambled next, and the crew last of all. The baby went up in a bag with the opening tied up . . . we set foot on deck with very thankful hearts.[8]

By a quarter after eight, all of the boats had snuggled up next to the *Carpathia* except Boat 12. Second Officer Lightoller's boat had seventy-four people and was still a quarter mile away and low in the water. She was moving slowly, as she was only designed to hold sixty-five people, and Lightoller did not want the boat to get swamped in the chop that had come with the breeze. The wind had kicked up, and whitecaps now appeared on the once-placid sea. According to Lightoller, "Every wave threatened to come over us. All were women and children in the boat apart from those of us men from the Englehart."[9]

Captain Rostron, seeing the difficulty of Boat 12, engaged his engines and moved the *Carpathia* forward and to starboard to bring Lightoller's boat into the ship's lee side. But a squall kicked up with the wind, and when Rostron turned, several waves went over the boat. "We couldn't last many minutes longer, and round the *Carpathia*'s bows was a scurry of wind and waves that looked like defeating all my efforts after all. One sea lopped over the bow, and the next one far worse. The following one she rode."[10] Lightoller then jammed his tiller over, and Boat 12 slid to the sheltered side of the ship and she was secured by 8:30.

Jack Thayer wrote later of the rescue:

There was a rope ladder with wooden steps hanging down her side. Most of us climbed up it, although many had to be hauled up in slings or chairs. . . . As I reached the top of the ladder, I suddenly saw my mother. When she saw me, she thought, of course, that my father was with me. . . . It was a terrible shock to hear that I had not seen Father since he had said goodbye to her.[11]

Lightoller began unloading his passengers and heard a familiar voice above him. "Hullo, Lights! What are you doing down there?"[12] Lightoller looked up and saw his old friend, First Officer Horace Dean, staring down from the side of the ship. He had been the best man at his wedding.

The passengers came on board with a backdrop of a pink sea of icebergs stretching out to the horizon. The lifeboats coming to the large ship in the middle of the Atlantic gave the setting a strange sense of community as the passengers came aboard in a serendipitous assortment of clothing. One woman had wrapped a Turkish towel around her waist and tossed a fur evening cape over her shoulders. The first-class passengers promenaded in: "lace trimmed evening dresses . . . kimonos . . . fur coats . . . plain woolen shawls . . . rubber boots . . . white satin slippers."[13] Many still had on hats one might expect to see in New York City, with large Edwardian hats for the women and some of the men in tweed caps.

Thayer found himself with a coffee cup full of brandy. "It was the first drink of an alcoholic beverage I had ever had," said Thayer.

"It warmed me as though I had put hot coals in my stomach. . . . A man kindly loaned me his pajamas and his bunk, then my wet clothes were taken to be dried, and with the help of the brandy I went to sleep till almost noon."[14]

The silence aboard the *Carpathia* was oppressive. There was the shock of the sinking, death on a massive scale as fifteen hundred people drowned within hearing of those in the boats. Beesley later wrote about boarding the *Carpathia*:

> *The passengers crowded the rails and looked down on us as we rowed up in the early morning, stood quietly aside while the crew at the galley ways took us aboard, and watched us as if the ship had been in dock. . . . Some of them have related that we were very quiet as came aboard. . . . There was very little excitement on either side, just the quiet demeanor of people who are in the presence of something too big as yet to lie in with their mental grasp, and which they cannot yet discuss.*[15]

Those in Lightoller's boat boarded the *Carpathia* with a mixture of emotions. Colonel Gracie, who had tried to revive a person who died next to him in the boat, felt like "falling down on his knees and kissing the deck."[16] Harold Bride, the surviving radio operator, felt strong hands pull him up as he passed out from exhaustion, cold, and severe frostbite in his feet.

Now all of the *Titanic's* survivors had been brought on board. The pursuer did a quick count, and by 9:30 a.m., it was known that the rescue had resulted in 705 people being plucked from the lifeboats on the Atlantic. This was both heartening and devastating. The math was brutal. More than fifteen hundred people had frozen to death or drowned in the icy water. But this was also Jack Phillips and Harold Bride's shining moment. Because of the new technology of wireless telegraphy and the dedication of the two men who had not stopped transmitting until the water was over their ankles, people had been saved. The 705 people on board the *Carpathia* did not understand that they owed their lives to the slight young man passed out in the first-class dining room.

And they needed his services again. Beesley, along with many of the *Titanic* passengers, wanted to let the world know right way they had survived.

> One of the first things we did was crowd round a steward with a bundle of telegraph forms. He was the bearer of the welcome news that passengers might send Marconi grams to their relatives free of charge, and soon he bore away the first sheaf of hastily scribbled messages to the operator.[17]

But there was only Cottam, the lone operator, and, as Beesley surmised, "The pile must have risen high in the Marconi cabin."[18] Most of the passenger messages would never be sent.

Interestingly, Beesley later wrote that when the *Carpathia* was sighted in his boat, one name was mentioned with the "deepest feeling of gratitude: that of Marconi. I wish that he had been there to hear the chorus of gratitude that went out to him for the wonderful invention that spared us many hours, and perhaps many days, of wandering about the sea in hunger and storm and cold."[19]

Jack Thayer woke up the next day "and got up feeling fit and well, just as though nothing bad had happened." He added, "After putting on my clothes, which were entirely dry, I hurried out to look for mother."[20] He found that Captain Rostron had given up his cabin to his mother, Mrs. George D. Widener, and Mrs. John Jacob Astor. Chivalry knew no bounds with Captain Rostron. Thayer noted later that when he saw Bruce Ismay, he was shocked. "I am almost certain that on *Titanic* his hair had been black with slight tinges of gray, but now his hair was virtually snow white."[21]

CHAPTER THIRTY-SEVEN

My God, They Are All Lost

Six Hours after the Titanic Sank

THE YOUNG MEN OF THE HOGDON WIRELESS CLUB CROWDED AROUND the wireless set in the laboratory of the high school building in Passaic, New Jersey. Engineer Brady of the high school had given them the set, and now they were listening as the Morse code was transcribed and messages from Newfoundland found their way into the laboratory. The boys heard the long and short buzzes that said the *Titanic* had sunk but there were people being rescued in boats.

Then they heard that the *Titanic* was being towed to Halifax. Then around 2 p.m. came a message that only twenty boats were rescued. The signals from the middle of the Atlantic Ocean lit up the faces of the teenagers who had no idea they were listening in real time to the biggest drama of the century. Those same signals were finding their way to the *Olympic*, which was still racing furiously toward the *Titanic* with every furnace fired up and every stoker shoveling coal as fast as possible into the fiery maul of the boilers. Even though it would seem that the passengers had been rescued, there was still some question as to the fate of *Titanic*'s passengers. Moore, the wireless operator on the *Olympic*, was worried. He had not heard from Jack Phillips for hours, and they had been old friends. Even though the official line from New York was that all passengers were safe, there were conflicting reports that made him doubt this.

Roy W. Howard, the general manager of United Press Association, was on board and wanted to send messages to his business partners on shore, but Moore could not diverge from trying to get updates on the *Titanic*. *New York Times* reporter W. Orton Teson was on board as well, and he reported later on the news that everyone was safe. Iceberg warnings came in from the *Parisian* and the *Asian*, and the *Olympic* had altered its course around the ice fields. Wireless operator Moore was getting frustrated by the lack of information concerning the *Titanic*. It was already 1 p.m. Monday and the signals from Cape Race were fuzzy. Other messages were being jammed by the German liner *Berlin*. The messages that did come through were requests from newspapers in New York, for example, "Will pay you liberally for story of rescue of *Titanic* passengers."[1] Some told Moore to name his price for any information he could provide. Moore replied to Cape Race, which was relaying the requests. "It's no use sending messages from newspapers asking us for news of *Titanic*, as we have none to give."[2] Finally, at 3:50 p.m. came a reply from *Carpathia* to *Olympic*'s inquiries, "Steady on. I don't do everything at once. Patience please,"[3] as Harold Cottam, the wireless operator, responded. Then came the news that no one wanted:

> *I received distress signals from the* Titanic *at 11:20, and we proceeded right to the spot mentioned. On arrival at daybreak, we saw field ice 25 miles, apparently solid, and a quantity of wreckage and number of boats full of people. We raised about 670 souls. The* Titanic *has sunk. She went down in about two hours. Captain and all engineers lost. Our captain sent order that there was no need for the* Baltic *to come any further. So, with that she returned course to Liverpool. Are you going to resume your course on that information? We have two or three officers aboard and the second Marconi operator who had been creeping his way through water 30 degrees sometime. Mr. Ismay aboard.*[4]

Operator Moore immediately relayed the message to the bridge and then forwarded it to Cape Race. The bulletin in the first-class dining and smoking room was discreetly taken down. There was no more obfuscation, there was no more hope. The *Titanic* had sunk, and only 705 people had survived. Moore sent back to Cottam, "Any urgent messages for shore we will send through Cape Race for you, but please stand by for service (ship to ship) messages." Cottam sent back. "OK Old Man, but I'm tired and hungry. Have had nothing to eat since 5:30 yesterday."[5]

It was Moore who handed the message to Captain Haddock on the bridge. Haddock knew many of the officers on the *Titanic* and just two weeks before he and Captain Smith had switched commands. It was unbelievable. The unsinkable ship had sunk, and the only survivors were in twenty lifeboats. Captain Haddock stared out from the bridge and couldn't imagine what those people had gone through. But here was the official confirmation that the world had dreaded, and the Cape Race Station didn't get the information to New York for two hours. In that two hours every conceivable rumor arose from the electromagnetic storm of signals crisscrossing the ocean.

Crazy messages attributed to the *Carpathia* came around 2 p.m. "All passengers of the *Titanic* safely transferred to the ship and SS *Parisian*. Sea Calm. *Titanic* being towed by Allen liner *Virginian* to port." The special train to Halifax was canceled. The conservative *New York Sun* ran a headline built from the wireless traffic: "All Saved from *Titanic* after collision."[6] The *Sun* went on to attack the *New York Times* for reporting that the *Titanic* had sunk. The information that the *Titanic* passengers were going to arrive in New York circulated and boosted IMM stock after wild fluctuations. But David Sarnoff, on top of Wanamaker's department store in New York, had been glued to his wireless set ever since he received the first information that the *Titanic* had struck an iceberg.

The twenty-one-year-old listened through the static and the jamming of amateur operators who were responsible for a lot of the disinformation. He could hear the very faint signals of the *Titanic*'s sister ship the *Olympic*. She was fourteen hundred miles out at sea, but the signals had bounced around and managed to hit the antenna strung up over the roof of Wanamaker's on the edge of Manhattan. It was the news that changed the world. "The *Titanic* had foundered at 12:47 a.m., New York time, and the only known survivors, about 675 people, were aboard the *Carpathia*, now bound for New York."[7] The news stopped the party of rumors that somehow said *Titanic* had survived, and it was replaced with the appalling news of mass death in 1912. To the Edwardian world of the early twentieth century, innocence was shattered with this news given by a twenty-one-year-old on top of a department store.

"I have often been asked what my emotions at that moment were," David Sarnoff later said.

I doubt if I felt at all during the seventy-two hours after the news came. I gave the information to the press associations and newspapers at once, and it was as if bedlam had let loose. Telephones were whirring, extras were being cried, crowds were gathering around newspaper bulletin boards.[8]

The dike of hope that Vice President Franklin had erected at the IMM offices collapsed with the news that there had been extreme loss of life. Still, Franklin, after admitting a "terrible loss of life," held out an ember of hope, "We are hopeful that the rumors which have reached us by telegraph from Halifax that there are passengers aboard the *Virginian* and the *Parisian* will prove to be true, and that these vessels will turn up some of the passengers," he said. "It is the loss of life that makes this thing so awful."[9]

When a confirming message came from Cape Race, Franklin broke down at midnight. "I thought her unsinkable," he sobbed, "and I based my opinion on the best expert advice. I do not understand it."[10] For many, the sinking of the *Titanic* was something they could not comprehend. By 1 a.m., a crowd of four thousand people had gathered in Times Square around the newspaper bulletin where a young man scribbled the latest news. The millionaires began to make their way to the White Star offices. Vincent Astor, the son of millionaire Jacob Astor, went into the office and emerged sobbing. A Mr. Hoyt, who had a brother and sister-in-law on the liner, also went into the office, only to emerge and confront the stricken faces of those seeking information. "My God," cried Hoyt. "They are all lost!"[11]

Novelist Theodore Dreiser, upon hearing the news onboard the ship *Kroonland*, went out and stared out into the dark night, leaning against the rail. He later wrote, "We went to the rail and looked out into the blackness ahead. The swish of the sea and the intermittent moan of the foghorn. . . . I went to my berth thinking of the pain and terrors of those doomed two thousand, a great rage in my heart against the fortuity of life."[12]

CHAPTER THIRTY-EIGHT

More Like an Old Fishing Boat Had Sunk

Seven Hours after the Titanic *Sank*

CAPTAIN ROSTRON STEAMED AROUND THE AREA LOOKING FOR ANY more survivors after the last lifeboat had been unloaded. Lawrence Beesley observed the *Carpathia* "cruising round and round the scene but found nothing to indicate there was any more hope of picking up more passengers. . . . There was surprisingly little wreckage to be seen: wooden deck chairs and small pieces of other wood, but nothing of any size."[1] Beesley noted a strange phenomenon of "huge patches of reddish yellow seaweed."[2] It was said be cork, but it had a strange, ghastly presence in the area where so many had died. More strangely, there were no corpses to be seen. The Atlantic current had already taken the floating cadavers away.

The *Carpathia* had been headed for Gibraltar but now the question was where to land the *Titanic* passengers. Captain Rostron thought about continuing on to Gibraltar, then to the Mediterranean to drop the passengers at the Azores where he could drop off the survivors to be picked up by another White Star Line ship. But Purser Brown pointed out they did not have the necessary food to continue on their original voyage. Rostron then considered the closest port, Halifax, but that would put him in line with a large ice field, and the captain considered his passengers had seen enough icebergs for a lifetime.

No. There was only one option. Turn the ship around and head back to New York. There would be some initial ice, but then it should be clear sailing and they could make good time on the return trip. Captain Rostron stopped down at Ismay's cabin and ran the plan by him. The now almost-catatonic manager of White Star mumbled his acquiescence and Rostron charted the course back to New York. Jack Thayer went down to visit Ismay during this time in his cabin after the doctor of the *Carpathia* suggested it might do the president of the White Star Line some good. It was a strange request and makes no real sense, but the youth's impression of Ismay right after the sinking is one of the few. "He was seated in his pajamas, on his bunk, straight ahead, shaking all over like a leaf. . . . I have never seen a man so completely wrecked. Nothing I could do or say brought any response."[3]

The *Carpathia*'s passengers did all they could for the *Titanic*'s passengers, giving them extra clothes, toiletries, toothbrushes. Some gave up their cabins or made room for passengers, while others served coffee and hot tea. But there were some on the *Carpathia* who were beyond consoling. When Mrs. Ogden brought over some coffee to some women in the corner of the upper deck, they waved her away, saying, "Go away . . . we've just seen our husbands drown."[4]

Captain Rostron dealt with the *Titanic*'s boats by bringing aboard thirteen of the lifeboats with six in the *Carpathia*'s davits and seven stowed on the foredeck. The rest were set adrift. While the boats were being loaded, the *Mount Temple* pulled up six miles to the east and Rostron filled her in and asked her to continue the search for any survivors. He went back to the chart room to continue plotting his course when he was called back to the bridge.

Operator Harold Cottam had a message from the *Olympic* from Captain Haddock requesting the transfer of passengers to his ship. Captain Rostron thought this would be bad for the passengers, seeing a twin of the *Titanic* pull up. He went down and checked with Bruce Ismay, who shook his head. Rostron had Cottam send back a message to the *Olympic* declining the offer. Now, at 9:15, Captain Rostron saw another ship, a four-masted freighter with a single red funnel. She seemed to appear out of nowhere, but First Officer Dean and Second Officer Bisset recognized the ship as the one they had seen off to the northwest. Rostron didn't know who the ship was and was puzzled, since Cottam had told him that aside from the *Mount Temple* no ship was close enough yet to render assistance. The ship signaled her identity, the *Californian*.

Captain Rostron requested Captain Lord to look for survivors and let him know he was heading back to New York. Before they set sail for New York, Rostron held a brief service for those who had perished and also to commit to the deep the bodies of the eight crew who had been taken out of the boats dead. Reverend Father Anderson, an Episcopal clergyman, held the service in the main lounge with the passengers from the *Carpathia* and the *Titanic*. While Father Anderson held the service, the *Carpathia* steamed over what was now the *Titanic*'s grave. The morning sun played off the clear ice of the giant bergs and mirrored the water. This was really the last moment for the *Titanic*.

Captain Rostron, who had risked everything to come to the aid of the *Titanic*, felt he had done everything he could and turned his ship toward New York, leaving the *Mount Temple* and the *Californian* to look for the dead. The *Carpathia* began to steam away, and Beesley, watching from the decks, observed the "sea covered with solid ice, white and dazzling in the sun and dotted here and there with icebergs." He continued, "We ran close up, only two or three hundred yards away, and steamed parallel to the floe, until it ended towards night and we saw to our satisfaction the last of the icebergs."[5]

Captain Rostron's ship now became a magnet for a world that wanted to know what had happened, and the race began to get back to New York with the survivors. The race started with *Parisian* radio operator Donald Sutherland, who woke up and detected a faint message at 7 a.m. from the *Carpathia* that said she had passengers from the *Titanic*. He went up to the bridge and told Captain Haines, who told him to find out anything else he could, and Haines began plotting a route back toward the *Titanic*. Sutherland learned from the *Virginian* that the *Carpathia* had all the survivors on board, and his captain decided not to turn around.

The *Baltic* was still heading for the *Titanic* when she made wireless contact sometime after 8 a.m. "Can I be of any assistance to you as regards to taking some of the passengers from you? Will be in position about 4:30 p.m. Let me know if you alter your position."[6] No reply came back, but then the *Baltic* tried again, and Cottam responded and explained they had the survivors on board and communicated Captain Rostron's message: "Am proceeding to New York top speed. You had better proceed to Liverpool. Have about 800 passengers on board."[7] The *Baltic* turned around after altering her course by 134 miles.

The crew of the *Mount Temple* still nosing around in the ice saw another ship steaming up at 8 a.m. It was the *Birma* coming from the southwest, churning up black smoke from her full-speed run. The Russian ship came to a halt at the coordinates of the *Titanic* sinking as given by Jack Phillips. Captain Stulping looked around, puzzled at the lack of wreckage. Stulping saw the *Mount Temple* and the *Californian* off to the east and began to weave through the ice. There were now three ships milling around the ice field when a fourth rolled up from the southwest. The German ship SS *Frankfurt* had come to the position of the sinking and steered east to get around the ice field.

The four ships cruised through the ice looking for survivors, bodies, debris. It was amazing. It was as if the *Titanic* and all her passengers had simply vanished. The race to rescue the *Titanic* had produced five ships that had managed to get to the site of the sinking but there was no one to rescue after the *Carpathia* departed. The *Mount Temple* contacted the *Carpathia* at 8:30 and Cottam sent back a message from Rostron saying all had been done and the *Mount Temple* should return to her course.

The *Frankfurt* listened in and heard the message that Captain Hattorff had ordered his crew to start stowing away mattresses, blankets, and first aid equipment. The men on the ships realized then that they were too late for most of the passengers on the *Titanic*. The *Virginian* also picked up the message and slowed down and turned back to her original course. The *Birma* also resumed her original course and headed off along with the *Frankfurt*. The *Virginian* got in touch with the *Carpathia*, and Rostron sent back the same message: "We are leaving here with all on board, about 800 passengers. Please return to your northern course."[8]

This left the *Californian* to search for survivors. Captain Lord continued maneuvering through the ice looking for any survivors or any bodies, for that matter. He saw only "some deck chairs and cushions." Lord later said he was surprised at the lack of wreckage, commenting, "It was more like an old fishing boat had sunk."[9] The captain of the *Califor-*

nian made a last circle around the debris and, just before noon, decided to head out. He had never admitted seeing the *Titanic* the night before, and the lack of wreckage made him feel better. He was irritated and felt he was on a fool's mission. Maybe the *Titanic* had been much farther away than people knew, and that would mean she was much farther away from the *Californian*. And that would mean his men did not see *Titanic* or her rockets. And to Captain Stanley Lord, that was more important than anything else.

CHAPTER THIRTY-NINE

We Must Get the Wireless Man's Story

One Day after the Titanic *Sank*

THE WORLD WANTED TO KNOW WHAT HAPPENED. IT WOULD TAKE three days for the *Carpathia* to reach New York, and the race was on to get the scoop on what actually happened to the *Titanic*. Now everyone knew it had sunk and the world was in shock. In England, J. P. Alexander of Manchester, a member of Parliament, read the morning headlines and dropped dead. C. P. Sumner was frantic. He was the general manager of the Cunard Line and a crowd of people had barged into his office demanding to know who had survived. Sumner had sent no less than five messages to Captain Rostron, but there had been no reply.

The entire world was now looking for the electromagnetic wave from a ship in the middle of the Atlantic Ocean that would change the lives of some 2,200 people. More than 1,500 people had died and there were 675 to 855 survivors, depending on what newspaper you read. The wait was excruciating. President William Howard Taft was one of those desperate for information. His friend and personal advisor Archie Butt had been on the ship after suffering a near nervous breakdown. Teddy Roosevelt had thrown his hat into the ring to run again for president, and Butt had been torn between his loyalties to the former and present president. The trip to Europe was to be recuperative; now Archie Butt was one of the many that might or might not be on the *Carpathia*.

President Taft couldn't take the suspense and ordered two cruisers, the *Chester* and the *Salem* to go the site of the sinking and make contract with the *Carpathia* and get a list of survivors—and if need be, bring Archie Butt home. The stock market fluctuated wildly. IMM stock plummeted and shares of Marconi stock shot through the roof. Information was king, but there was none forthcoming from the *Carpathia* at the moment. So, every radio operator in America scanned the airwaves looking for any information. Wireless telegraphy was the social media of its time, and, like the personal computer, many people had their own set to listen in to other people and send out their own broadcasts.

In Kansas, government operator at Fort Leavenworth, Private Boyer, listened for messages from the northeast. Reporters sat beside him with headphones and waited to write down any news he might receive. The wind howled outside the wireless station with rain pattering the windowpanes. A fire burned cheerily in the fireplace. Boyer chewed on a cigar and listened intently and then picked up a pen. The reporters leaned in eagerly as Boyer shook his head and dropped the pen. He turned the knob of his serial condenser to try to get the message again, touching it with his pen. "No use," Boyer exclaimed. "He's quit. It was the Wanamaker station in New York. The static is something awful tonight." There had been a strange amount of static in the air all week. As one newspaper reported, "Static electricity . . . has greatly interfered with wireless telegraphy all over the country for a week or more, and the records at the *Post* station show that very little government business has been conducted by wireless as a result."[1]

Boyer adjusted his condenser and then his detector and coils, and a message came in. "It's San Antonio,"[2] said Boyer, and disconnecting his detector he pushed a button. "From a neighboring room came a hum that increased to a shriek as the transformers stepped up the city voltage from 110 volts to 2,500 volts, and then the operator seized the sending key and, with a tremendous snapping and humming of blue sparks, he sent an answer."[3] Boyer finished and said it was nothing but the latest news from the *San Antonio Post* but there was nothing regarding the *Titanic*. The reporters slumped back down to wait.

At the *Brooklyn Daily Eagle*, the newsroom was held hostage to the man with a harness over his head holding two receivers to his ears. He would listen, then pause and write down what had just come in, and then go back to listening at his desk. The *Eagle* wireless station was trying, along with everyone else, to get news from *Carpathia*. Every minute, ships and stations were sending out the call letters MPA for the *Carpathia*, but nothing was coming back at all. The Brooklyn Navy Yard station was trying to get in contact with the cruiser *Salem*, which had made contact with the *Carpathia*, but now there was nothing.

The *Carpathia* had transmitted an initial list of survivors and then gone silent. Newspapers ran headlines of frustration: "*Carpathia* Lets No Secrets of the *Titanic* Loss Escape by Wireless."[4] Even President Taft could get no response from the ship of survivors. But initial lists of survivors had been transmitted, and the first-class passengers got the news first. A reporter phoned John Jacob Astor IV's father-in-law, W. H. Force, for a reaction to Astor's death. "Oh my God," cried Force. "Don't tell me that. Where did you get that report from? It can't be true!"[5]

Mrs. Alfred Hess, the daughter of Macy's founder Isidor Straus, had taken the train to Halifax when, outside of Maine, the train stopped and began to run backward and didn't stop until Boston, where she received a message. "Plans have changed; the *Titanic* people are going straight to New York."[6] She took the train back the city and was met by her brother, who murmured, "Things look pretty bad."[7] With the first survivor list came crowds storming the White Star office. Mrs. Benjamin Guggenheim gained entrance and pleaded with White Star officials to launch a rescue mission, as she believed her husband was in a lifeboat somewhere. "He may be drifting about!" she pleaded.[8] The newspapers, having no real information about the sinking, made up their own stories. "The *Evening World* told of a fog, the *Titanic*'s booming siren, a crash like an earthquake. The *Herald* described how the ship was torn asunder, plunged into darkness, and almost capsized at the moment of impact."[9] No one knew, and the *Carpathia* was not telling. Nefarious motives began to be ascribed to the ship's silence. Someone was censoring the information that should have been forthcoming. Who would have thought that the very censor of information was the man responsible for bringing wireless communication to the world and whose technology had saved the 705 people that were on the *Carpathia*? He had struck a match against the darkness of the North Atlantic, and now Guglielmo Marconi was snuffing that same match out.

The Marconi operator on the SS *Florida* had picked up several messages received by the *Carpathia* as she approached New York. They were signed by Marconi himself. By Thursday, April 18, the anticipation of the rescue ship's arrival was intense. Reporters from throughout the world had been sent to cover the arrival, and the lack of information beyond the passenger lists had ratcheted up the drama to a fever pitch. The temperature was turned up even more when an official inquiry from the president of the United States was refused in trying to ascertain the status of advisor Archibald Butt.

The *San Francisco Examiner* ran a headline: "Taft Wire Refused by Rescuing Vessel."[10] The reason for the *Carpathia*'s silence seemed to be found in the intercepted messages meant for Harold Bride, the *Titanic* wireless operator:

> *Say old man. Marconi Company taking good care of you. Keep your mouth shut. It's fixed for you to get good money. Arranged for your exclusive story for dollars in four figures. Mr. Marconi agreeing. Say nothing until you see me. Where are you now?*
>
> *Go to Strand Hotel 582 West Fourteenth Street and see Mr. Marconi.*[11]

Marconi knew this was the story of the century, and he wanted to control the narrative. Bride was the man who could tell the world what happened, and even with the agreement between Marconi and the *Times*, Arthur Greaves, the paper's city editor, was taking no chances. He pulled in all his reporters the day of the *Carpathia*'s arrival and told them the following:

> *The* Carpathia, *with* Titanic *survivors, is due tonight around 9 o'clock. She has not answered wireless requests for information. A.P. just sent us this note. "We have no assurance that we will get any wireless news from the* Carpathia, *as this vessel studiously refuses to answer all queries. . . ." I'm sending sixteen of you down to the pier, though we only have four passes. Men without passes will have to try and get through—to survivors, crew, and passengers—on their police*

cards. . . . Get all you can. Get Captain Rostron of the Carpathia.
Get Bruce Ismay of the White Star. Get every possible member of the
Titanic's *crew, especially the four officers who were saved. We must get
the* Titanic's *wireless man's story, if he's alive.*[12]

Frustration at the lack of information and the inability to communicate with the *Carpathia* grew to such a point on Thursday, April 18, that the amateur radio operators all along the East Coast were blamed. A blanket order agreed to by the Marconi Company and the navy said that all wireless stations would be shut down except for a designated few. Since everyone was on the same frequency, the shutting down of amateurs, as well as government stations, was an attempt to clear the airwaves. It didn't help. As the *Carpathia* came into view on the rainy, foggy night of April 18, 1912, only Marconi knew the reason for the lack of information, and it wasn't the amateur operators.

CHAPTER FORTY

Completely Destitute, No Clothes

Two Days after the **Titanic Sank**

THE SHIP THAT HAD THE ONLY SURVIVORS OF THE *TITANIC* ON BOARD was making her way through stormy seas. The bow rose and fell with waves crashing across her decks. For the *Titanic* passengers this was insult to injury. They were ill clothed, sleeping in makeshift quarters, and now the ocean that had taken the *Titanic* was tossing the smaller *Carpathia* all around. The survivors were slumped in deck chairs now or holding warm cups of coffee in the dining room.

A storm on Wednesday night caused a panic among the survivors sleeping in the library. A crack of lighting was so intense and the thunder so loud that many third-class passengers thought the ship was in trouble and women grabbed their children and dashed onto the decks as they had on the *Titanic*. Order was restored and the *Carpathia's* passengers did whatever they could for the distressed survivors, giving them tooth-brushes or clothes or stitching together smocks from *Titanic* steamer blankets for the children.

Lawrence Beesley wrote about the extra effort the *Carpathia* passen-gers made for the survivors. They were now

> *hard at work finding clothing for the survivors, the barber shop was raided for ties, collars, hair pins, combs. . . . One good Samaritan went round the ship with a box of toothbrushes offering them indiscrimi-nately to all. In some cases, clothing could not be found for the ladies,*

and they spent the rest of the time in their dressing gowns and cloaks. . . . Women had to sleep on the floor of the saloons and in the library each night on straw paillasses. . . . The men were given the smoking room and a supply of blankets.[1]

Beesley managed to get off a message that he was safe and then fell into a dreamless sleep, utterly exhausted from the ordeal. Colonel Gracie was buried under a pile of blankets trying to bring warmth back to his body in the dining saloon while his clothes lay draped over the baker's oven. Bruce Ismay had furtive messages sent to White Star that they must try to get himself and the entire *Titanic* crew onto a ship bound for England when they arrived. He did not want to face any inquiries into the United States that might ask him how it was he had survived when so many others had perished.

Wireless operator Harold Bride had passed out when he came aboard the *Carpathia* and was taken to a stateroom, where he came to "lying in somebody's stateroom, a woman was bending over him, and he felt her hand brushing back his hair and rubbing his face."[2] Captain Rostron directed Harold Cottam to keep the channel open and to ignore press inquiries. "He instructed Cottam to give first priority to communicating news to Cunard and White Star offices, in particular sending the names of the survivors, then passing along any brief personal messages they might have. Messages sent by the *Carpathia*'s own passengers would have to wait."[3]

At 3 p.m., Cottam sent Captain Rostron's message to Captain Haddock of the *Olympic.* "Fear absolutely no hope searching *Titanic*'s position. Left Leyland SS *Californian* searching around."[4] He then gave a summary of information he had gleaned from the surviving officers of the *Titanic* and he gave an exact number of the survivors he had on board. This was when Haddock suggested that the survivors be transferred to the *Olympic*, and, after consulting with Ismay, Rostron sent back, "Mr. Ismay's orders. *Olympic* not to be seen by the *Carpathia*. No transfer to take place."[5] Captain Haddock then inquired if he should continue to the scene of the sinking. Rostron said there were no survivors except those on his ship. Haddock then slowed down and turned around, heading back toward Europe. This left the *Carpathia* as the lone ship of the *Titanic* rescue mission.

Then the *Carpathia* came to a stop as Rostron decided to bury at sea the bodies he had on board. The bodies were sewn into a canvas covering with a weight at their feet. "The engines were stopped, and all passengers on deck bared their heads while a short service was read."[6] Operator Moore on the *Olympic* began to assist Cottam in handling radio traffic and relaying messages with its much more powerful wireless transmitter. All press inquiries were ignored, with priority given to lists of survivors, yet Cottam was hopelessly overwhelmed. The lists were daunting enough, but then there were personal missives from survivors to relatives, and Cottam had not slept for more than twenty-four hours, with little to eat. About 4:15 p.m. the radio transmitter on the *Carpathia* stopped sending. The passenger lists being transmitted to the *Olympic* and shore stations stopped abruptly. The ships that had been trying to contact her nonstop heard nothing at all.

A steward brought some tea to the wireless room and found Cottam slumped over his equipment with his hand still on the telegraph key. He had been on duty for more than forty hours, and his body had given out. A doctor diagnosed "utter exhaustion." Cottam was carried to his bunk where he slept for the next twenty-four hours. Piled-up telegrams and the much more important survivor lists were on Cottam's desk. Captain Rostron needed an operator and no one on board was trained in wireless telegraphy. That left Harold Bride, the *Titanic* operator who was down in third class wrapped in a blanket with his feet heavily bandaged against the severe frostbite he had incurred on Collapsible D. Bride knew Cottam. The two men were friends and, more than that, brothers in arms in the world of telegraphic operators on ships. Bride volunteered to take over for Cottam and he hobbled to the wireless room where he put on the headphones and began to tap out to the world again. The man who had been among the last on the *Titanic*, along with Jack Phillips, transmitting to the end, was back on the airwaves again, letting the world know of the carnage he had seen.

Captain Rostron had Bride send a message to the *Olympic* to be relayed to New York: "Will send names immediately ready, you can understand we are working under considerable difficulty. Everything possible being done for comfort of survivors."[7] Rostron sent a message to Cunard in New York. "*Titanic* struck iceberg Monday 3 a.m. 41.46 N, 50.14 W. *Carpathia* picked up many passengers in boats will wire further particulars later proceeding back to New York."[8]

The entire world was waiting for news, and by Monday night Cottam and Bride had given the *Olympic* the list of survivors of first and second class, followed by third class and crew. Cottam had listened in on the *Olympic*'s transmissions to double-check the names. Then the wireless cabin on the *Carpathia* was besieged with passengers who wanted to let loved ones know they had survived. The *Minnewaska* reported, "*Carpathia* will not send traffic to Sable Island and insists on sending lists of names to me, saying they are for our captain. . . . Operator on *Carpathia* evidently tired out."[9]

Then Sable Island operators late on Tuesday evening tried to contact the *Carpathia*, but the airwaves were so jammed with the inquires being sent that nothing could get through. Even Ismay's official confirmation of the disaster would not be sent until Wednesday, April 17. The entire world was ready to tar and feather Harold Cottam and Harold Bride as the news blackout continued, but Cottam pointed out later that one reason Ismay's message was not sent until Wednesday was that "it was policy not to send company messages to shore via ships of other lines, and as such he had waited until *Carpathia* was in range of Sable Island." When Captain Rostron was asked about the delay, he said Ismay was "mentally very ill," but the message was taken by the purser to Cottam with his permission to send as soon as possible. Rostron continued, "I did not forbid relaying message to any ship. On contrary, I particularly mentioned doing all possible to get official messages, names of survivors, then survivors' messages away by most convenient means."[10]

The silence of the *Carpathia* annoyed the U.S. government so much that President Taft sent the USS *Chester* to find out what was going on by contacting the *Carpathia*. So hard did the USS *Chester* harry the *Carpathia* for information that the Siasconset shore station found the airwaves blocked. With the navy operators holding down the Morse key for ten and thirty seconds at a time, Siasconset's operator reported the system paralyzed on account of the navy. Bride must have felt harassed at this point. After surviving the sinking of the *Titanic*, he had the entire world on the same frequency trying to get information, with the navy ship deliberately tying up the airwaves in a sort of wireless filibuster. The Siasconset operator on Nantucket Island, Massachusetts, was frustrated as well, and by Wednesday night he had more than a hundred messages for the *Carpathia* that he could not send.

The operator tried for four hours, and then, on Thursday morning, he contacted the navy operator, who told him he had an important message for the *Carpathia*, "but he won't take it. This seems to be the case for our own Siasconset traffic," the shore operator fumed.

Over 110 (messages) on hand, yet we have been unable to raise the Carpathia *since 9:50 p.m., and 2:09 a.m., when after attracting his attention he disappears, ignores our calls, and offers of messages, important messages, etc. Atmospherics are quite bad, but traffic can be worked off through it, especially with the good signals we are offering* Carpathia. *Looks like a case of laziness, don't care or something of that sort.*[11]

It is a stunning accusation by the Siasconset operator. Bride had stayed on the *Titanic* long after Captain Smith had given the order of every man for himself, with the water literally coming up to his knees while the wireless room approached a forty-five-degree angle toward the sea. Yet, he and Phillips continued to send out their signals until the power failed in the hope that some ship might pick them up that was close. If the power had not failed, Bride and Phillips would have gone down with the ship, there is no doubt.

In addition to having catastrophic frostbite and suffering from shock, hypothermia, and sheer exhaustion, Bride had to deal with a roomful of frantic passengers and stacks of unsent messages because another operator had literally collapsed over his key with his finger pressed. The world may have wanted information, but technology had not yet caught up with the appetite, and wireless telegraphy was still agonizingly slow. Still, Bride would later give the reason he did not communicate with the USS *Chester* to a newspaper, stating, "They were wretched operators. They knew American Morse but not Continental Morse sufficiently to be worthwhile. They taxed our endurance to the limit."[12]

Wireless telegraphy was so new that it was not unlike the early days of personal computers where only programmers really knew what was going on, and there was a certain elitism among the operators. The anger of Phillips when the *Californian* operator broke in so loudly it startled him happened a lot. Also, Marconi was in competition with other wireless companies, and this contributed to the reluctance to deal with competing Morse systems and to not communicate with ships with competing equipment.

The Siasconset operator asked if his signals could be read. "Yes, can read you OK, but say OM. I have not been to bed since *Titanic* went down. I have over 300 messages." This was Thursday and *Titanic* sunk on Monday morning. Incredibly, Harold Bride had not been to bed for three days at this point. Still, the Siasconset operator showed no mercy. "Make a try and less argument."[13]

Then Marconi sent a message at 3 p.m. "Wire news dispatches immediately to Siasconset or to navy boats if this is possible. Ask captain the reason why no news allowed to be transmitted. Guglielmo Marconi."[14] This was a disingenuous attempt to show Marconi was not sitting on the news from his operators. Five hours later, he sent another message. "Marconi officers . . . arranged for your exclusive story for dollars in four figures. Mr. Marconi agreeing. Stop. Say nothing until you see me."[15] Clearly he was pleased with the stonewalling of information, but Bride, in the *New York Times* story, would justify his silence, explaining, "In the first place, the public should not blame anybody because more wireless messages from the disaster to the *Titanic* did not reach the shore from the *Carpathia*. I positively refused to send press dispatches because the bulk of personal messages with touching words of grief was so large."[16]

Bride and Cottam were doing two things at once. They were in the business of letting people know who had died and who had lived, while fielding ridiculous offers of money for any information they would give. A message received from the *Boston Globe* was typical:

Anxious to get full narrative disaster can you write it for delivery to globe representative on dock New York will pay liberally rush answer to Winfield Thompson globe man on steamer Franconia *his expense he also desires wireless story of your experiences for immediate transmittal Boston every moment precious intense interest at home.*[17]

Bride was being offered more money than he would probably see in a lifetime of being a wireless operator while sending out messages of grief and heartbreak. These messages had to be exceedingly difficult to send for Bride, who had seen more than a thousand people perish in the water.

Father not seen no hope Arrive *Carpathia* Wednesday.

Every boat watched Father Mother not on *Carpathia* . . . hope still.

The ship sank, BM and I are safe.

Saved not Tyrell yet.

Mrs. Hays Mrs. Davidson safe no news of husband.

Titanic gone safe on *Carpathia.*

Safe on board *Carpathia* fear Joe lost.

Holrey lost. Meet me. Mate.

Sir Cosmo and Lady Duff-Gordon saved.

Mollie, Elizabeth, Natalie, Caroline safe on *Carpathia.* George Not heard from.

Jack Margaret I safe. No news Johnny.

Am safe. Pray God George was rescued by another boat with rest of men.

Let any one meet us but not children. My hope is gone.

Titanic Sunk! Saved on Board. Cunard Line *Carpathia* Completely Destitute, no clothes. Alfred.[18]

Captain Rostron held several meetings with Second Officer Lightoller on the tragedy, but he really wanted to find out from Bruce Ismay, the chairman of the White Star Line, what had happened. Ismay seemed in no condition to talk to anyone. The *Carpathia* now was getting closer to New York with her engines running open but nowhere near the speed she had run to get to the *Titanic*. Lightoller and the purser listed the survivors and came up with the numbers that would haunt the world. There were 2,223 people on the *Titanic*, and only 705 had been rescued by the *Carpathia*. Sixty percent of first class had been saved, 42 percent of second class, and 24 percent of third class. One hundred percent of children in second class had been saved, with only 23 percent of the crew. Seventy-five percent of the women and children had survived, with only twenty percent of the men. A total of 1,517 people had frozen to death or drowned in the North Atlantic.

The numbers were brought to the bridge, and they shocked Captain Rostron, a deeply religious man. He wondered if he might have done anything else that would have allowed him to get to the *Titanic* sooner. The only way he could had reached her in time is if he had been closer. A ship ten or even twenty miles away could have arrived in plenty of time. So many ships came to the rescue of the *Titanic*; it was hard to believe that none were closer than the fifty-eight miles he had needed to traverse. A ship that was ten miles away could have saved the fifteen hundred people who froze to death. Another thing that had been nagging the captain of the *Carpathia* was the amount of people in the lifeboats. Many were only half full, and yet none of the boats had returned to pick up more.

Captain Rostron stared off the bridge of the *Carpathia* into the fathomless sea and shook his head. He believed someone else had had their hand on the helm as he weaved through the icebergs at full speed. Why else was his ship the only one to get through and rescue the *Titanic* passengers in the lifeboats? Surely the other captains all acted as he did and turned about at full speed when they learned of her distress. No. No one would sit by and let people die in the freezing waters of the Atlantic. He had to believe that. During the journey, the *Carpathia* encountered "every kind of climatic conditions, icebergs, ice fields, and bitter cold to commence with; brilliant warm sun, thunder and lightning in the middle of one-night, cold winds most of the time, fogs every morning and during a good part of the day."[19]

The four days aboard the *Carpathia* became tedious. Lawrence Beesley noted that many felt the effects of the "ship crowded far beyond limits of comfort, the want of necessities of clothing and toilet."[20] Incredibly, few were ill besides Ismay's shattered mental health and Bride's severely frostbitten feet. When land finally came into sight, the ship was "surrounded by tugs of every kind, from which magnesium bombs were shot off by photographers, while reporters shouted for news of the disaster and photographs of passengers."[21] The *Carpathia* drew to her station at the Cunard pier, and, "We set foot at last on American soil, very thankful, grateful people," related Beesley.[22]

Lightning flashes and a thunderstorm were the backdrop on the scene of thousands waiting for the ship that would tell the story of the sinking of the RMS *Titanic*. A man in a dark derby with other men trailing went aboard and directly to the wireless shack. Bride was still bent over his telegraphic key sending the personal messages for passengers when Guglielmo Marconi entered the wireless shack and finally relieved the last wireless operator of the *Titanic*.

"That's hardly worth sending now, boy."[23]

Harold Bride looked up. His eyes flared, recognizing the famous inventor, who quickly grasped his hand and held it warmly.

CHAPTER FORTY-ONE

Endings

TITANIC MARCONI OPERATOR HAROLD BRIDE FINISHED TALKING AND the men in the hotel room were silent. Reverence. Shock. Stupefaction. Respect. How does a human being react after hearing the tale of the biggest disaster of the twentieth century? There simply were no words. There is a picture of Bride being carried off the *Carpathia*. He has on a fedora, an overcoat, and both his feet are bound up in bandages. He looks like a boy in bad spy movie with the hat pulled low and his arms around the shoulders of the two men carrying him. We can only assume Marconi was just behind or in front of the photographer. The surviving Marconi operator gave his story to Jim Speers of the *New York Times* and it filled up five columns of the front page of the *Times*.

It was the action-packed sea story everyone had been waiting for. The story begins with Bride asleep when the captain enters the wireless room and the breaking down of the wireless set before. "The lucky thing was that the wireless broke down early enough for us to fix it before the accident. We noticed something wrong on Sunday, and Phillips and I worked seven hours to find it. We found a secretary burned out, at last, and repaired it just a few hours before the iceberg struck."[1] This is often overlooked in the story of the *Titanic*, but had the wireless still been not functioning, everyone aboard the ship would have surely perished. Bride had just gotten up to relieve Jack Phillips when the captain stuck his head in the door. "We've struck an iceberg. And I'm having an inspection done. . . . You better get ready to send out a call for assistance."[2] Then he describes Phillips sending the CQD, followed by Bride suggesting they use the SOS as well. Then came Bride's description of the ship sinking by the bow and how the numbers of boats were not even enough for half the passengers. He was the first to say how the band played and how he and Phillips realized no ship would reach them in time. Then he described the final moments with the wave that swept him overboard.

There were men all around me—hundreds of them. The sea was dotted with them, all depending on their life belts. I felt simply I had to get away from the ship. She was a beautiful sight then. Smoke and sparks were rushing out of her funnel, and there must have been an explosion, but we had heard none. We only saw the big stream of sparks . . . the band was still playing, and I guess they all went down. They were playing "Autumn" then. I swam with all my might. I suppose I was 150 feet away when the Titanic, *on her nose, with her after quarter sticking straight up in the air, began to settle—slowly. When at last the waves washed over her rudder there wasn't the least bit of suction I could feel.*[3]

Bride went on to defend himself against charges he had not sent news when requested to the *Chester*, saying, "The navy operators were a great nuisance. I advise them all to learn the Continental Morse and learn to speed up if they ever expect to be worth their salt."[4] He describes his ten hours in the hospital on the *Carpathia* until he was requested to assist.

I took the key, and I never left the wireless cabin after that. Our meals were brought to us. . . . We worked all the time. Nothing went wrong. Sometimes the Carpathia *man sent, and sometimes I sent. There was a bed in the wireless cabin. I could sit on it and rest my feet while sending sometimes.*[5]

It is in this article that Harold says he saw Jack Phillips. As he was getting out of the lifeboat to climb up to *Carpathia*, he saw Phillips. "One man was dead," said Harold. "I passed him and went to the ladder, although my feet pained me terribly. The dead man was Phillips. He had been all in from work before the wreck came."[6] It is the first time we find out what happened to Phillips. Bride said that on the overturned collapsible he was exhausted and in a state of semiconsciousness. When he left the overturned boat for another lifeboat, this is where he claimed to have noticed Phillips lying dead in the boat. No one else mentioned Phillips being on board. It is one of the unanswered questions of the *Titanic* drama: What happened to the heroic Phillips in the final moments of her sinking? But if Phillips was on the boat and passed away, with Bride seeing him at the last moment, at least Bride knew what had happened to his friend.

He then described for readers being on the overturned collapsible. "There was just room for me to roll on the edge. I lay there not caring what happened. Somebody sat on my legs. They were wedged in between the slats. . . . At first the larger waves splashed over my clothing. Then they began to splash over my head, and I had to breathe when I could."[7] When the *Carpathia* pulled up, Bride managed to climb the rope ladder and then passed out. He woke up in a cabin. "The next thing I knew, a woman was leaning over me in a cabin, and I felt her hand waving back my hair and rubbing my face. I felt somebody at my feet and felt the warmth of a jolt of liquor."[8]

The front-page article made Harold Bride world famous and enshrined Jack Phillips in the pantheon of heroes for all time. Bride's description of Phillips at his post as the *Titanic* sunk was immortalized in song and verse. Said Bride, "He was a brave man. I learned to love him that night, and I suddenly felt for him a great reverence to see him standing there sticking to his work while everybody else was raging about. I will never live to forget the work of Phillips for the last awful fifteen minutes."[9]

After Bride gave his story, he was taken to St. Vincent's Hospital and then to the home of a relative in New York on West 92nd Street. Newspapers reported that "his nervous condition is such that he will be compelled to spend several weeks in uninterrupted rest."[10] It was also reported that his salary would be paid until he had recovered. His nervous condition was undoubtedly shock.

Bride would give testimony at the American and British inquiries. In the American inquiry he was questioned as to why he did not respond to the U.S. Navy inquires, and it was revealed that Marconi had arranged the story with the *New York Times* and the order to keep quiet. Bride replied as he had in the newspapers that the U.S. Navy did not understand British Morse signals and that his priority was survivors lists and personal messages. The matter was dropped, and Bride, along with Marconi for inventing wireless telegraphy, was singled out as one of the heroes of the *Titanic's* sinking.

Two days after the *Carpathia* arrived in New York, the *Californian* docked in Boston. There was no crowd to greet her, but a representative from the Leyland Line went aboard immediately and locked himself away with Captain Lord in his cabin. Before this, on April 18, Captain Lord had called in Second Officer Stone and Apprentice Officer Gibson to his cabin and had each man write a sworn statement pertaining to the events of the morning of April 15. He then put both statements in his safe.

Even with sworn statements, Captain Lord could not stop the rumors that immediately began flying around the waterfront that the *Californian* had been close enough to see the distress rockets of *Titanic* and to see the moribund liner sink. Boston reporters soon began popping up and asking to talk to Captain Lord. The *Boston Traveler*, *Boston Evening Transcript*, and *Boston Globe* all sent reporters to interview Lord. The captain immediately cast doubt on anyone who gave a different version of events. He then declined to show any reporters his log, saying his position was "state secrets."[11] It died down quickly, as the reporters took Lord at his word. But on April 23, 1912, the *Clinton* (Massachusetts) *Daily* ran a headline that put Captain Lord on the hot seat: "*Californian* Refused Aid—Former Carpenter On Board This Boat Says Hundreds Might Have Been Saved from *Titanic*."[12] James McGregor, the ship's carpenter, had told his cousin that the *Californian* was close enough to see the sinking *Titanic* and had seen her distress rockets. He went on to say the officers on watch had seen the rockets and had reported them to Captain Lord, who refused to take action. McGregor felt such bitterness toward the captain, he stated, that he "would positively refuse to sail under him again and that all the officers had the same feeling."[13] He was soon backed up by Ernest Gill, one of the assistant engineers aboard the *Californian*, and four other engineers who swore out a lengthy affidavit that purported the following:

> *He personally had seen a ship firing rockets just after midnight on April 15th. Moreover, he claimed that the ship firing the rockets was no more than ten miles away and that he had heard the* Californian's *second officer saying that he too had seen the rockets, and that the ship's captain had been told about them.*[14]

The *Boston American* ran the affidavit and sent a copy to Senator Smith in New York who was holding a Senate inquiry into the disaster. Captain Lord was subpoenaed, along with Gill, and both men testified on April 25. It was revealed that Gill was paid $500 for his story. Gill finished by declaring, "I am of the general opinion that the crew is confident that she was the *Titanic*."[15]

Captain Lord replied to these reports by saying the ship that his officers saw was not the *Titanic* because it was too small and didn't appear to be in any danger. It was a different ship that had stopped ten miles from the *Californian* just before midnight and then steamed away at half past two. He then debunked all the passengers, officers, and crew that claimed to have seen the *Titanic* or some ship that was on the horizon not ten miles away. He then said the rockets had only been reported to him once and they were not distress rockets. When questioned about his position in the ice relative to the ship that was firing the rockets, Lord said it would have taken him at least two hours to reach the ship. He then danced around the issue of the rockets.

"Captain, did you see any signals of distress on Sunday night, either rockets or the Morse signals?"

"No sir. I did not. The officer on watch saw some signals, but he said they were not distress signals."

"They were not distress signals?"

"Not distress signals."

"But he reported them?"

"To me. . . . I think you had better let me tell that story."[16]

Captain Lord then gave his rendition of events, ending with the claim that he wasn't quite awake when Apprentice Gibson had come down a second time to tell him of the rockets and the steamship steaming away. Senator Smith then asked him a simple question.

"Captain, these Morse signals are a sort of language or method by which ships speak to each other?"

"Yes sir. At midnight."

"The rockets that are used are for the same purpose and are understood, are they not, among mariners?"

"As being distress rockets?'

"Yes."

"Oh yes, you never mistake a distress rocket."[17]

Senator Smith was buying none of it. In his conclusion of the investigation, he wrote,

> *I am well aware from the testimony of the captain of the* Californian *that he deluded himself with the idea that there was a ship between the* Titanic *and the* Californian, *but there was no ship seen there at daybreak, and no intervening rockets were seen by anyone on the* Titanic—*although they were looking longingly for such a sign*—*and saw only the white light of the* Californian, *which was flashed the moment the ship struck and was taken down when the vessel sank. A ship . . . could not have gone west without passing the* Californian *on the north or the* Titanic *on the south. That ice floe held but two ships*—*the* Titanic *and the* Californian.[18]

Senator Smith then shredded Captain Lord's veracity for all time by condemning him for failing to come to the assistance of the *Titanic*, "in accordance with the dictates of humanity, international usage, and the requirements of law."[19] This conclusion was mild compared to the scathing conclusion of the British investigation into the sinking that was headed by Lord Mersey.

The truth of the matter is plain. The Titanic *collided with the berg at 11:40. The vessel seen by the* Californian *stopped at this time. The rockets sent up from the* Titanic *were distress signals. The Californian saw distress signals. The number sent up by the* Titanic *was about eight. The* Californian *saw eight. The time over which the rockets from the* Titanic *were sent up was from about 12:45 to 1:45 o'clock. It was about the same time that the* Californian *saw the rockets. At 2:40, Mr. Stone called to the Master that the ship from which he's seen the rockets disappeared.*

At 2:20 a.m., the Titanic *had foundered. It was suggested that the rockets seen by the* Californian *were from some other ship, not the* Titanic. *But no other ship to fit this theory has ever been heard of. These circumstances convince me that that the ship seen by the* Californian *was the* Titanic, *and if so, according to Captain Lord, the two vessels were about five miles apart at the time of the disaster. . . . The evidence from the* Titanic *corroborates this estimate, but I am advised that the distance was probably greater, though not more than eight to ten miles. . . . When she first saw the rockets, the* Californian *could have pushed through the ice to the open water without any serious risk and so have come to the assistance of the* Titanic. *Had she done so, she might have saved many if not all of the lives that were lost.*[20]

Captain Lord's reputation was shattered, and he was relieved of his command by the Leyland Line. He was eventually hired in 1913 by the Nitrate Producers Steamship Company, where he stayed until 1927 when he retired, citing health issues. He mounted an effort to clear his name in 1958, but died in 1962 before his petition was rejected in 1965. In 1968, his son launched another petition to clear his name, which was also rejected. Books both supporting Lord and convicting him have been coming out ever since, along with movies that generally portray him in a harsh light. The discovery of the *Titanic* on the ocean floor reignited the controversy.

Captain Moore of the *Mount Temple* was also called to testify in both inquires. When questioned about his reluctance to cross the ice field, Moore responded,

My instructions from my company are that I must not enter field ice,
no matter if it seems only light. Those are the explicit instructions from
my company. If I was to go through the ice and my ship was damaged,
I would have pointed out to me that those were the instructions, that
I was not to go into any ice, no matter how thin.[21]

But the passenger and crew witnesses who claimed to have seen the
Titanic's rockets would not go away. It was when she reached Canada that
the crew and passengers spoke to the press and said they had seen the
rockets launched from the *Titanic*. This has led to speculation about how
far the *Mount Temple* was from the *Titanic* when she received her call for
help, how far she was when she stopped, and Captain Moore's decision
not to cross the ice field. The speculation that Moore ignored *Titanic*'s
rockets has been the basis for several books and many articles.

In 2009, Senan Molony wrote the book Titanic *Scandal: The Trial of
the* Mount Temple, where he postulates that the *Mount Temple* was in fact
the mystery ship that Captain Smith sent off two of his lifeboats in pur-
suit of. He also points out that the ship pursued by these lifeboats slowly
turned away, which matches Captain Moore's decision to not broach the
ice field, even though, according to passengers and crew, he could see the
Titanic firing rockets in a desperate plea for help. As compared to Captain Lord, Moore's career was left relatively intact, even though questions
persist to this day.

Captain Rostron of the *Carpathia* testified at both the U.S. Senate
inquiry and the British inquiry. At the American inquiry he made a deep
impression on Senator Smith, who chaired the proceedings. He "had
instantly impressed everyone with his courage, his clear headedness, his
thoroughness, and his compassion." Moreover,

Rostron's dignified and disciplined bearing affected everyone on the
committee, and when he described the memorial service held aboard the
Carpathia, *as well as the funeral for four survivors who died shortly*
after they were rescued, there were tears in his eyes. At the obvious sor-
row of this sunburnt seamen, many in the hearing openly wept.[22]

It quickly became clear that Captain Rostron was the yardstick by which all others would be judged, for the "alacrity with which Rostron had responded to the *Titanic*, the comprehensive preparation he made, and the courage he had shown by steaming full speed into the ice field to pick up survivors."[23] At the British inquiry, Captain Rostron surprised the investigators when he said he and another office saw the lights of another ship when he arrived to rescue the *Titanic* passengers. When questioned he answered, "I saw the masthead lights."

"Did you see the lights your officer spoke of?"

"I saw the masthead light myself, but not the sidelight."

"What time was it?"

"About a quarter past three."

"And how was the light bearing?"

"About two points on the starboard bow."

"On your starboard bow?"

"On my starboard bow, that would be about N.30, W. true."[24]

The revelation was damning for Captain Lord. Rostron was saying he had visual sight of another ship while he was at the site where *Titanic* sunk. In other words, as he was picking up the lifeboats, there was another ship that was "clearly within visual distance of the spot where the *Titanic* sunk."[25] This was the *Californian*, and after Rostron's incredible fifty-eight-mile run, he had to wonder why this ship had not come to the aid of the *Titanic*. The reason she did not come to her aid was that Captain Lord had yet to wake up, and this left the operators of the *Californian* to observe the second ship sending off green flares, which was the *Carpathia*.

Captain Rostron held the inquiry in thrall as he described once again how he had pushed his ship, "approaching the icefield, dodging icebergs, relying on the sharp eyes of his officers, his own skill as a seaman, and his faith in divine intervention."[26] Rostron was lauded for his actions and ultimately awarded the Congressional Gold Medal, the Thanks of Congress, the American Cross of Honor, a medal from the Liverpool Shipwreck and Humane Society, and a gold medal from the Shipwreck Society of New York. He was highly praised by governments around the world, but the silver cup and gold medal given to him for his efforts to rescue the passengers of the *Titanic* on the night of April 14, 1912, meant more to Captain Rostron than all the rest. It was from the survivors.

Captain Arthur Henry Rostron returned to the command of the *Carpathia* before taking command of the *Caronia*. He then took command of the *Lusitania*, *Campania*, and *Carmania* from 1913 to 1914. During World War I, he was captain of the *Aulania* and was involved in the battle of Gallipoli. He took command of the RMS *Mauretania* in 1915, and the following year he joined the *Ivernia* before returning to the *Mauretania* a year later. At the war's end in 1919, he was made a commander of the Order of the British Empire.

Captain Rostron continued on with the *Mauretania* until 1924, serving as Royal Naval Reserve aide-de-camp to King George V. In 1926, he was knighted as a knight commander of the Order of the British Empire and took command of the RMS *Berengaria* and later became the commodore of the Cunard fleet. He retired in 1931, and wrote his autobiography, *Home from the Sea*. He died November 7, 1940, survived by his wife, Ethel Minnie, and their four children.

CHAPTER FORTY-TWO

The Race to Rescue the *Titanic*

THE RACE TO RESCUE THE *TITANIC* BEGINS AND ENDS IN THAT TILTING, sepia-lit wireless room with water sloshing around the ankles of two men furiously tapping out calls for assistance in the middle of the North Atlantic. Those electromagnetic signals draping every antenna of every ship and every shore station asked a simple question: Will you come to our assistance? CQD. SOS. It was all the same. Will you come to our assistance? Because we are sinking and will die if you do not.

This question was asked of every person who received those signals, and they all then had 160 minutes to get to the RMS *Titanic* before she sank and took more than fifteen hundred people with her. When all the books and all the studies and all the theories and movies and discussions are stripped away, there is this one question at the core of the rescue operation on the night of April 14, 1912. Will you come rescue us before we perish? And how the ships, people, companies, nations handled this simple request is the story of the *Titanic*. The mythology would have us believe that her sinking was preordained. That in reality there was no way the people who drowned on the *Titanic* could have been saved, and so the story became the doomed tragic sinking of the ship with all the pathos that implies. The heroics of the Greek tragedy were played out under the assumed doom of the exceptional liner and gave great import to the stories of the band playing on and Edwardians prepared to die like gentlemen, saving women and children first.

The heroic white male was never better than on the night of April 14, 1912, with his sense of noblesse oblige, bidding farewell to wives, daughters, sweethearts, while watching from the railing of the doomed liner with a last shot of brandy, a final cigar, a last hardy handshake before going down into the Atlantic. This story, then, is the one that made for great books and even greater movies. It is a hell of a story. But the real story is very different. In fact, it is the opposite. The *Titanic* was not doomed to take her passengers with her. In fact, most of the people—if not all—might have been rescued were it not for *the failing of the heroic white males* in their hour of need. The very invention of the wireless puts the lie to the heroic story we have been handed of WASP moral superiority in the dire moment. The story used the wireless to augment its fable, but the truth is the wireless *put the test to three men, and two of them came out wanting*.

The messages from the *Titanic* put the moral dilemma front and center for Captain Rostron, Captain Lord, and Captain Moore, and for the first time, these captains were powerless to shape the narrative. Before wireless, there was no one to contest the captain's will. No one. Out on the North Atlantic only one man dictated reality and that was the captain. The wireless broke this iron fist and presented the outside world's judgment against the immaculate rule of order on the high seas. So the last thing Stanley Lord wanted to do was wake up his wireless operator on the night of April 14 when his officers watched the *Titanic* fire off rockets of distress and then sink in clear view. If he had woken the wireless man and received the information that the *Titanic* was sinking right in front of him, *he would have been compelled to enter the ice field and rescue the drowning passengers*. Ten miles away. Maybe less. Maybe more, depending on whose book you pick up, but the truth is that Captain Lord was presented a dilemma, because his wireless operator told him when he asked what ship was in the vicinity, and only the *Titanic* was the answer. And to wake up that same operator would puncture Lord's grip on events that night. He would lose the ability to dance away from the facts. With the wireless operator asleep, the old days of the ship isolated and dependent on her captain for all information was intact, and Captain Lord could shape the narrative. What undid him in the end was that he was so close that many on his ship saw the *Titanic* and then told the press. His iron grip on the facts was ultimately destroyed by his officers and crew.

Captain Moore on the *Mount Temple* was the closest, save for the *Californian*. She received those distress signals and immediately headed for the *Titanic*. But then, with the *Titanic* in sight, according to the crew and passengers who saw the sinking ship and later told newspapers, she turned around. It was not a few passengers who saw the *Titanic* and not a few crewmen. Captain Moore had decided the ice was not worth the risk to himself and his passengers. It was in his DNA to be cautious, and like everyone else who received that call for help he had a decision to make. Tragically, he was found wanting. And even though Captain Moore denies ever seeing the *Titanic*, he does admit that he stopped the ship because of the ice field and would not risk his ship. Seeing the *Titanic* in the distance only makes the optics worse, and, according to the crew who considered mutiny, they could not persuade him to go forth even though the boats were lowered and ready to go.

Captain Rostron of the *Carpathia* who received the message from Harold Bride and Jack Phillips orchestrated the rescue mission of the *Titanic*. The wireless's new role on the seas was honored, and he turned his ship around and risked all to get to the *Titanic* within the 160 minutes. Captain Rostron was too far away and reached the site one hour and twenty minutes after the *Titanic* sank, but he saved 705 people, which makes the race to rescue the *Titanic* a partial success. The final undoing of the *Titanic* mythology comes from the passengers themselves. Many of those in the lifeboats were first-class passengers, the cream of society, the inheritors of noblesse oblige. Steerage or third-class passengers were loaded only when the first class had been safely sent off into the ocean. And yet this smorgasbord of the top rung of society decided on the whole to *not go back and help the people drowning right in front of their boats*. This would be the final breakdown of moral high ground for the "Teutonic people," as Lawrence Beesley called them that night. The WASP credo of bravery, chivalry, and honor was shredded on the night the *Titanic* went nose first into the North Atlantic. It doesn't matter that the band played on, they played on not for the heroic, but for the cowardice of white privilege that essentially, in its moment of truth, shouted its real clarion call. And it was not pretty.

The mythology of the *Titanic* is one of self-sacrifice, but the reality is one of *self-preservation*. Ayn Rand's novel *Atlas Shrugged*'s basic premise that altruism or acting out of anything other than self-preservation goes against human nature seems to hold up here. Out of the twenty boats, only one returned that night to pick up people in the water, and by then it was mostly too late, only rescuing four people. The rest of the *Titanic*'s finest sat in their boats in their dresses, tuxedoes, and furs and tried to drown out the chorus of cries and the death wail so many would carry into their dreams. Many sang. Some prayed. The primal urge to survive is a reality, and one must recognize this, but it should not be covered up with a Gilded Age patina.

The tragedy is compounded by the fact that many of the lifeboats were only half full and there was room in total for about four hundred more people. That would have left eleven hundred in the water to be picked up by the two ships so close that some lifeboats rowed toward their lights. But this was not to be. The heroes of the *Titanic*, wireless operators Jack Phillips and Harold Bride, could only ask for help. They needed the character of Captain Rostron to be on the receiving end of their electromagnetic waves, but they had only Captain Lord of the *Californian* who did not deign to even get up when he was told there was a ship firing distress rockets, and Captain Moore of the *Mount Temple* who turned back from the very ice field the *Titanic* was trapped in when the sinking ship was in sight. Sadly, on the night of August 14, 1912, courage was found lacking when it was needed most.

The race to rescue the *Titanic* involved nine ships that came to her aid. They were all brought by the miracle of wireless telegraphy that for the first time would give the world a real-time window into an unspeakable tragedy as it was unfolding. Imperfect as it was, for the first time, ordinary citizens sitting at their kitchen tables could grasp what was happening in the middle of the frozen Atlantic. From David Sarnoff to the high school boys who picked up signals from the *Titanic*, *Olympic*, and the Cape Race Station, the veil of black silence had been broken and now the world could visualize what was happening to the largest ship ever built as it sank within a scant 160 minutes. The heroism, the cowardice, the finest moments, and the worst moments were now on display. That is what the race to rescue the *Titanic* showed in the microcosm of three different captains. Conspiracy theories aside, and there are many on all sides, three different men reacted in three different ways, and against this backdrop were the captains of the ships that did come thundering to the rescue, even though they were much too far away.

We would like to think that we would react as Captain Rostron did, weaving our way through icebergs, putting every ounce of steam into our engines, going through the night with the only purpose being that of rescuing those in distress. Or that if we were in the lifeboats and saw the fifteen hundred people drowning and our lifeboat was half full, we would demand to go back and risk getting swamped to save those we could. This is the human ideal, and it is almost impossible to know if we would turn back from the ice, as Captain Moore did, or refuse to recognize the ship plainly in view firing distress rockets, as Captain Lord did. The truth is that human beings are a blend, and that is why we all sit back and doff our hats to the hero. They may be rarer than we like to think.

But there were real heroes that night, and we are once again back to those two wireless operators Phillips and Bride tapping away and sending out their basic question again. It is a universal question really. A very human cry that comes down through the years and still tests us every time we hear it. On the *Titanic* lying on the ocean floor, the wireless set from 1912 remains, and it is fitting that, as of this writing, there is a plan for it to be brought back to the surface by underwater robots. Many feel this will be a fitting tribute to Harold Bride and Jack Phillips, and perhaps it will have one last message embedded in its transmitter coils. Undoubtedly, if they could extract that last plea, that last electromagnetic impulse, it would be that very human question we still all ask of one another: Will you come help us?

CHAPTER FORTY-THREE

"Thrilling Story by *Titanic*'s Surviving Wireless Man"

New York Times
April 19, 1912
Harold Bride

Bride Tells How He and Phillips Worked and How He Finished a Stoker Who Tried to Steal Phillips's Life Belt—Ship Sank to Tune of "Autumn"
(THIS STATEMENT WAS DICTATED BY MR. BRIDE TO A REPORTER FOR THE NEW York Times *who visited him with Mr. Marconi in the wireless cabin of the* Carpathia *a few minutes after the steamship touched her pier.*)

In the first place, the public should not blame anybody because more wireless messages about the disaster to the *Titanic* did not reach shore from the *Carpathia*. I positively refused to send press dispatches because the bulk of personal messages with touching words of grief was so large. The wireless operators aboard the *Chester* got all they asked for. And they were wretched operators.

They knew American Morse but not Continental Morse sufficiently to be worthwhile. They taxed our endurance to the limit. I had to cut them out at last they were so insufferably slow, and go ahead with our messages of grief to relatives. We sent 119 personal messages today and 50 yesterday. When I was dragged aboard the *Carpathia*, I went to the hospital at first. I stayed there for ten hours. Then somebody brought word that the *Carpathia* wireless operator was "getting queer" from the work.

They asked me if I could go up and help. I could not walk. Both my feet were broken or something, I don't know what. I went up on crutches with somebody helping me. I took the key, and I never left the wireless cabin after that. Our meals were brought to us. We kept the wireless working all the time. The navy operators were a great nuisance. I advise them all to learn the Continental Morse and learn to speed up in it if they ever expect to be worth their salt. The *Chester*'s man thought he knew it, but he was as slow as Christmas coming.

We worked all the time. Nothing went wrong. Sometimes the *Carpathia* man sent, and sometimes I sent. There was a bed in the wireless cabin. I could sit on it and rest my feet while sending sometimes. To begin at the beginning, . . . I joined the *Titanic* at Belfast.

I didn't have much to do aboard the *Titanic* except to relieve Phillips from midnight until sometime in the morning, when he should be through sleeping. On the night of the accident, I was not sending, but I was asleep. I was due to be up and relieve Phillips earlier than usual. And that reminds me—if it hadn't been for a lucky thing, we never could have sent any call for help.

The lucky thing was that the wireless broke down early enough for us to fix it before the accident. We noticed something went wrong on Sunday, and Phillips and I worked seven hours to find it. We found a "secretary" burned out, at last, and repaired it just a few hours before the iceberg was struck. Phillips said to me as he took the night shift, "You turn in boy, get some sleep, and go up as soon as you can and give me a chance. I'm all done for with this work of making repairs."

There were three rooms in the wireless cabin. One was a sleeping room, one a dynamo room, and one in an operating room. I took off my clothes and went to sleep in bed. Then I was conscious of waking up and hearing Phillips sending to Cape Race. I read what he was sending. It was traffic matter. I remembered how tired he was, and I got out of my bed without my clothes on to relieve him. I didn't even feel the shock. I hardly knew it had happened after the Captain came to us. There was no jolt whatever. I was standing by Phillips telling him to go to bed when the Captain put his head in the cabin.

"We've struck an iceberg," the Captain said. "and I'm having an inspection made to tell what it has done for us. You'd better get ready to send out a call for assistance. But don't send it until I tell you." The Captain went away, and in ten minutes, I should estimate the time, he came back. We could hear a terrible confusion outside, but there was not the least thing to indicate that there was any trouble. The wireless was working perfectly. "Send the call for assistance," ordered the Captain, barely putting his head in the door.

"What call should I send?" Phillips asked.

"The regulation international call for help. Just that."

Then the Captain was gone. Phillips began to send, "C.Q.D."

He flashed away at it, and we joked up while we did so. All of us made light of the disaster.

We joked that way while he flashed signals for about five minutes. Then the Captain came back.

"What are you sending?" he asked.

"C.Q.D," Phillips replied.

The humor of the situation appealed to me. I cut in with a little remark that made us laugh, including the Captain. "Send S.O.S," I said, "It's the new call, and it may be your last chance to send it."

Phillips, with a laugh, changed the signal to "S.O.S." The Captain told us we had been struck amidships, or just back of amidships. It was ten minutes, Phillips told me, after he had noticed the iceberg, that the slight jolt that was the collision's only signal to us occurred. We thought we were a good distance away.

We said lots of funny things to each other in the next few minutes. We picked up first the steamship, *Frankfurt*. We gave her our position and said we had struck an iceberg and needed assistance. The *Frankfurt* operator went away to tell his captain. He came back, and we told him we were sinking by the head. By that time we could observe a distinct list forward. The *Carpathia* answered our signal. We told her our position and said we were sinking by the head. The operator went to tell the captain and in five minutes returned and told us that the captain of the *Carpathia* was putting about and heading for us.

Our Captain had left us at this time, and Phillips told me to run and tell him what the *Carpathia* had answered. I did so, and I went through an awful mass of people to his cabin. The decks were full of scrambling men and women. I saw no fighting, but I heard tell of it. I came back and heard Phillips giving the *Carpathia* fuller directions. Phillips told me to put on my clothes. Until that moment I forgot that I was not dressed. I went to my cabin and dressed. I brought an overcoat to Phillips. It was very cold. I slipped the overcoat upon him while he worked. Every few minutes Phillips would send me to the Captain with little messages. They were merely telling how the *Carpathia* was coming our way and gave her speed.

I noticed as I came back from one trip that they were putting off women and children in lifeboats. I noticed that the list forward was increasing. Phillips told me the wireless was growing weaker. The Captain came and told us our engine room was taking water and that the dynamos might not last much longer. We sent that word to the *Carpathia*. I went out on the deck and looked around. The water was pretty close up to the boat deck. There was a scramble aft, and how poor Phillips worked though it I don't know. He was a brave man. I learned to love him that night, and I suddenly felt for him a great reverence to see him standing there sticking to his work while everybody else was raging about. I will never live to forget the work of Phillips for the last awful fifteen minutes.

I thought it was about time to look about and see if there was anything detached that would float. I remembered that every member of the crew had a special life belt and ought to know where it was. I remembered mine was under my bunk. I went and got it. Then I thought how cold the water was. I remembered I had some boots, and I put those on, and an extra jacket, and I put that on. I saw Phillips standing out there still sending away, giving the *Carpathia* details of just how we were doing. We picked up the *Olympic* and told her we were sinking by the head and were all about all down. As Phillips was sending the message, I strapped his life belt to his back. I had already put on his overcoat.

I wondered if I could get him into his boots. He suggested with a sort of laugh that I look out and see if all the people were off in the boats or if any boats were left or how things were. I saw a collapsible boat near a funnel and went over to it. Twelve men were trying to boost it down to the boat deck. They were having an awful time. It was the last boat left. I looked at it longingly a few minutes. Then I gave them a hand, and over she went. They all started to scramble in on the boat deck, and I walked back to Phillips. I said the last raft had gone. Then came the Captain's voice: "Men you have done your full duty. You can do no more. Abandon your cabin. Now it's every man for yourself. You look out for yourselves. I release you. That's the way of it at this kind of a time. Every man for himself."

I looked out. The boat deck was awash. Phillips clung on sending and sending. He clung on for about ten minutes or maybe fifteen minutes after the Captain had released him. The water was then coming into our cabin. While he worked something happened I hate to tell about. I was back in my room getting Phillips's money for him, and as I looked out the door I saw a stoker, or somebody from below decks, leaning over Phillips from behind. He was too busy to notice what the man was doing. The man was slipping the life belt off Phillips's back. He was a big man too. As you can see, I am very small. I don't know what it was I got hold of. I remembered in a flash the way Phillips had clung on—how I had to fix that life belt in place because he was too busy to do it.

I knew that man from below decks had his own life belt and should have known where to get it. I suddenly felt a passion not to let that man die a decent sailor's death. I wished he might have stretched rope or walked a plank. I did my duty. I hope I finished him. I don't know. We left him on the cabin floor of the wireless room, and he was not moving.

From aft came the tunes of the band. It was a ragtime tune; I don't know what. Then there was "Autumn." Phillips ran aft, and that was the last I ever saw of him alive. I went to the place I had seen the collapsible boat on the boat deck, and to my surprise I saw the boat and the men still trying to push it off. I guess there wasn't a sailor in the crowd. They couldn't do it. I went up to them and was just lending a hand when a large wave came awash of the deck. The big wave carried the boat off. I had hold of an oarlock, and I went off with it. The next I knew I was in the boat. But that was not all. I was in the boat, and the boat was upside down, and I was under it. And I remember realizing I was wet through and that whatever happened I must not breathe, for I was under water.

I knew I had to fight for it, and I did. How I got out from under the boat I do not know, but I felt a breath of air at last. There were men all around me—hundreds of them. The sea was dotted with them, all depending on their life belts. I felt I simply had to get away from the ship. She was a beautiful sight then. Smoke and sparks were rushing out of her funnel. There must have been an explosion, but we had heard none. We only saw the big stream of sparks. The ship was gradually turning on her nose—just like a duck that goes down for a dive. I had only one thing on my mind—to get away from the suction. The band was still playing. I guess all of the band went down.

They were playing "Autumn" then. I swam with all my might. I suppose I was 150 feet away when the *Titanic* on her nose, with her after quarter sticking straight up in the air, began to settle slowly.

When at last the waves rushed over her rudder there wasn't the least bit of suction I could feel. She must have kept going just so slowly as she had been. I forgot to mention that, besides the *Olympic* and *Carpathia*, we spoke to some German boat. I don't know which and told them how we were. We also spoke the *Baltic*. I remembered those things as I began to figure what ships would be coming toward us. I felt after a little while like sinking. I was very cold. I saw a boat of some kind near me and put all my strength into an effort to swim to it.

It was hard work. I was all done when a hand reached out from the boat and pulled me aboard. It was our same collapsible. The same crowd was on it. There was just room for me to roll on the edge. I lay there not caring what happened. Somebody sat on my legs. They were wedged in between slats and were being wrenched. I had not the heart to ask the man to move. It was a terrible sight all around—men swimming and sinking. . . .

At first the larger waves splashed over my clothing. Then they began to splash over my head, and I had to breathe when I could. As we floated around on our capsized boat and I kept straining my eyes for a ship's lights, somebody said, "Don't the rest of you think we ought to pray?" The man who made the suggestion asked what the religion of the others was. Each man called out his religion. One was a Catholic, one a Methodist, one a Presbyterian.

It was decided the most appropriate prayer for all was the Lord's Prayer. We spoke it over in chorus with the man who first suggested that we pray as the leader. Some splendid people saved us. They had a right-side-up boat, and it was full to its capacity. Yet, they came to us and loaded us all into it. I saw some lights off in the distance and knew a steamship was coming to our aid. I didn't care what happened. I just lay and gasped when I could and felt the pain in my feet. At last the *Carpathia* was alongside, and the people were being taken up a rope ladder. One boat drew near, and one by one the men were taken off of it.

One man was dead. I passed him and went to the ladder though my feet pained terribly. The dead man was Phillips. He had died on the raft from exposure and cold, I guess. He had been all in from work before the wreck came. He stood his ground until the crisis had passed, and then he had collapsed, I guess.

But I hardly thought that then, I didn't think much of anything. I tried the rope ladder. My feet pained terribly, but I got out to the top and felt hands reaching out to me. The next I knew a woman was leaning over me in a cabin, and I felt her hand waving back my hair and rubbing my face. I felt somebody at my feet and the warm jolt of liquor. Somebody got me under the arms. Then I was hustled down below to the hospital. That was early in the day I guess. I lay in the hospital until near night, and they told me the *Carpathia*'s wireless man was getting queer and would I help. After that I was never out of the wireless room. So I don't know what happened among the passengers. I saw nothing of Mrs. Astor or any of them. I just worked wireless. The splutter never died down. I knew it soothed the hurt and felt like a tie to the world of friends and home.

How could I then take news queries? Sometimes I let a newspaper ask a question and get a long string of stuff asking for full particulars about everything. Whenever I started to take such a message I thought of the poor people waiting for their messages to go—hoping for answers to them. I shut off the inquiries and sent my personal messages. And I feel I did the right thing. If the *Chester* had had a decent operator I could have worked with him longer, but he got terribly on my nerves with his insufferable incompetence. I was still sending my personal messages when Mr. Marconi and the *Times* reporter arrived to ask that I prepare this statement.

There were maybe one hundred left. I would like to send them all because I knew I could rest easier if I knew all those messages had gone to the friends waiting for them. But an ambulance man is waiting with a stretcher, and I guess I have to go with him. I hope my legs get better soon. The way the band kept playing was a noble thing. I heard it first when still we were working wireless when there was a ragtime tune for us, and the last I saw of the band when I was floating in the sea with my life belt on, it was still on the deck playing "Autumn." How they ever did I cannot imagine. That and the way Phillips kept sending after the Captain told him his life was his own and to look out for himself are two things that stand out in my mind over the rest.

Notes

TITANIC MYTHOLOGY

1. Veronica Hinke, *The Last Night on the* Titanic*: Unsinkable Drinking, Dining, and Style* (Washington, DC: Regnery, 2019), 14.
2. Richard Howells, *The Myth of the* Titanic (New York: Palgrave Macmillan, 1999), 99.
3. John Hamer, *RMS* Olympic (Raleigh, NC: Lulu, 2013), 262.
4. *New York Times*, April 19, 1912.
5. *New York Times*, April 19, 1912.
6. *New York Times*, April 19, 1912.
7. Hinke, *The Last Night on the* Titanic, 259.
8. Hinke, *The Last Night on the* Titanic, 2.
9. Walter Lord, *A Night to Remember* (New York: Henry Holt and Company, 1955), 72.

PROLOGUE

1. Marconi Wireless Telegraph Company, *Marconigraph*, vol. 1: 9.

CHAPTER ONE

1. Logan Marshall, ed., *On Board the* Titanic (Mineola, NY: Dover Publications, 2012), 121.
2. Marshall, *On Board the* Titanic, 121.
3. Wyn Craig Wade, *The* Titanic*: End of a Dream* (New York: Rawson, Wade, 1979), 48.
4. Wade, *The* Titanic, 48.
5. Daniel Butler, *Unsinkable: The Full Story of the RMS* Titanic (Mechanicsburg, PA: Stackpole Books, 1998), 170.
6. United States Congress, *"Titanic" Disaster: Hearings before a Subcommittee of the Committee on Commerce, United States Senate, Sixty-Second Congress, Second Session, Pursuant to S. Res. 283, Directing the Committee on Commerce to Investigate the Causes Leading to the Wreck of the White Star Liner "Titanic"* (Washington, DC: U.S. Government Printing Office, 1912), 483.
7. Kenneth L. Cannon II, "Isaac Russell's Remarkable Interview with Harold Bride, Sole Surviving Wireless Operator from the *Titanic*," *Utah Historical Quarterly*, vol. 81, no. 4 (Fall 2013), 334.
8. Cannon, "Isaac Russell's Remarkable Interview," 335.

9. Marc Raboy, *Marconi: The Man Who Networked the World* (Oxford, UK: Oxford University Press, 2016), 359.

10. Wade, *The* Titanic, 50.

11. Wade, *The* Titanic, 50.

12. Cannon, "Isaac Russell's Remarkable Interview," 335.

13. Wade, *The* Titanic, 51.

14. Cannon, "Isaac Russell's Remarkable Interview," 329.

15. Cannon, "Isaac Russell's Remarkable Interview," 339.

16. Harold Bride, "Thrilling Story by *Titanic*'s Surviving Wireless Man," *New York Times*, April 19, 1912.

CHAPTER TWO

1. Lawrence Beesley, *The Loss of the S.S.* Titanic (Boston: Houghton Mifflin, 1912), 31.

2. Beesley, *The Loss of the S.S.* Titanic, 31.

3. Beesley, *The Loss of the S.S.* Titanic, 41.

4. Beesley, *The Loss of the S.S.* Titanic, 71.

5. Beesley, *The Loss of the S.S.* Titanic, 71.

6. Beesley, *The Loss of the S.S.* Titanic, 44.

7. Beesley, *The Loss of the S.S.* Titanic, 23.

8. Beesley, *The Loss of the S.S.* Titanic, 44.

9. Lamar Underwood, *The Greatest Disaster Stories Ever Told* (Guilford, CT: Globe Pequot, 2003), 201.

10. Underwood, *The Greatest Disaster Stories Ever Told*, 201.

11. William Hoffman and Jack Grimm, *Beyond Reach: The Search for the* Titanic (New York: Beaufort Books, 2008), 52.

12. Terri Dougherty, Sean Stewart Price, and Sean McCollum, *Eyewitness to* Titanic (North Mankato, MN: Capstone, 2015), 64.

13. Sean Coughlan, Titanic*: The Final Messages from a Stricken Ship* (London: BBC News), 2012.

14. Stephanie Barczewski, Titanic*: 100th Anniversary Edition: A Night Remembered* (New York: Continuum, 2011), 12.

15. Veronica Hinke, *The Last Night on the* Titanic*: Unsinkable Drinking, Dining, and Style* (Washington, DC: Regnery, 2019), xiv.

16. John Eaton, Titanic*: Triumph and Tragedy* (New York: W. W. Norton, 1995), 113.

17. John Welshman, Titanic*: The Last Night of a Small Town* (Oxford, UK: Oxford University Press, 2012), 87.

18. Welshman, Titanic*: The Last Night of a Small Town*, 87.

19. Daniel Gaetán-Beltrán, ed., *The* Titanic (Farmington Hills, MI: Greenhaven Press, 2015), 33.

20. Robert Elliott Allinson, *Saving Human Lives: Lessons in Management Ethics* (New York: Springer, 2005), 89.

21. Samuel Halpern, "Navigational Inconsistencies of the SS *Californian*," http://www .titanicology.com/Californian/Navigational_Incosistencies.pdf.

22. Hinke, *The Last Night on the* Titanic, 21.

23. Hinke, *The Last Night on the* Titanic, 27.

24. Andrew Wilson, *Shadow of the* Titanic (New York: Atria Books, 2012), 257.

25. Tom Kuntz, *The* Titanic *Disaster Hearings* (New York: Pocket Books, 1998), 326.

26. Wyn Craig Wade, *The* Titanic: *End of a Dream* (New York: Rawson, Wade, 1979), 236.

CHAPTER THREE

1. Veronica Hinke, *The Last Night on the* Titanic: *Unsinkable Drinking, Dining, and Style* (Washington, DC: Regnery, 2019), xiii.

2. John B. Thayer, *The Sinking of the SS* Titanic, 2nd ed. (Chicago: Chicago Review Press, 2005), 328.

3. Thayer, *The Sinking of the SS* Titanic, 328.

4. Wyn Craig Wade, *The* Titanic: *End of a Dream* (New York: Rawson, Wade, 1979), 9.

5. Thayer, *The Sinking of the S.S.* Titanic, 334.

6. Jack Winocour, ed., *The Story of the* Titanic *as Told by Its Survivors* (New York: Dover Publications, 1960), 27.

7. Lamar Underwood, *The Greatest Disaster Stories Ever Told* (Guilford, CT: Globe Pequot Press, 2003), 402.

8. Nic Compton, Titanic *on Trial* (New York: Bloomsbury, 2012), 69.

9. Compton, Titanic *on Trial*, 70.

10. Compton, Titanic *on Trial*, 74.

11. Walter Lord, *A Night to Remember* (New York: Henry Holt and Company, 1955), 17.

12. Underwood, *The Greatest Disaster Stories Ever Told*, 402.

13. Compton, Titanic *on Trial*, 11.

14. Walter Lord, *The Night Lives On* (New York: William Morrow, 1986), 59.

15. Hinke, *The Last Night on the* Titanic, 177.

16. United States Congress, *"*Titanic*" Disaster: Hearings before a Subcommittee of the Committee on Commerce, United States Senate, Sixty-Second Congress, Second Session, Pursuant to S. Res. 283, Directing the Committee on Commerce to Investigate the Causes Leading to the Wreck of the White Star Liner "*Titanic*"* (Washington, DC: U.S. Government Printing Office, 1912), 1,028.

17. Lord, *A Night to Remember*, 7.

18. Lord, *A Night to Remember*, 18.

19. Lord, *A Night to Remember*, 18.

20. Gareth Russell, *The Ship of Dreams: The Sinking of the* Titanic *and the End of the Edwardian Era* (New York: Atria Books, 2019), 198.

21. Russell, *The Ship of Dreams*, 199.

22. Russell, *The Ship of Dreams*, 166.

23. Lord, *A Night to Remember*, 21.

24. Russell, *The Ship of Dreams*, 200.

25. Shan Bullock, *A* Titanic *Hero* (Riverside, CT: 7 C's Press, 1973), 69.

26. Russell, *The Ship of Dreams*, 200.

27. Daniel Butler, *Unsinkable: The Full Story of the RMS* Titanic (Mechanicsburg, PA: Stackpole Books, 1998), 48.

28. Andrew Wilson, *Shadow of the* Titanic (New York: Atria Books, 2012), 222.

29. Nick Barratt, *Lost Voices from the* Titanic (New York: St. Martin's, 2010), 130.

30. Michael Davie, *The* Titanic: *The Full Story of a Tragedy* (London: Bodley Head, 1986), 137.

31. Davie, *The* Titanic, 137.

32. Daniel Butler, *The Other Side of the Night: The* Carpathia, *the* Californian, *and the Night the* Titanic *Was Lost* (New York: Casemate, 2009), 63.

CHAPTER FOUR

1. Daniel Butler, *The Other Side of the Night: The* Carpathia, *the* Californian, *and the Night the* Titanic *Was Lost* (New York: Casemate, 2009), 51.

2. Butler, *The Other Side of the Night*, 49.

3. Butler, *The Other Side of the Night*, 49.

4. Nick Barratt, *Lost Voices from the* Titanic (New York: St. Martin's, 2010), 89.

5. Butler, *The Other Side of the Night*, 53.

6. Butler, *The Other Side of the Night*, 53.

7. Butler, *The Other Side of the Night*, 53.

8. Walter Lord, *The Night Lives On* (New York: William Morrow, 1986), 136.

9. Wyn Craig Wade, *The* Titanic: *End of a Dream* (New York: Rawson, Wade, 1979), 233.

10. Butler, *The Other Side of the Night*, 55.

11. Butler, *The Other Side of the Night*, 55.

12. Tom Kuntz, *The* Titanic *Disaster Hearings* (New York: Pocket Books, 1998), 326.

13. Butler, *The Other Side of the Night*, 56.

CHAPTER FIVE

1. Lawrence Beesley, *The Loss of the S.S.* Titanic (Boston: Houghton Mifflin, 1912), 53.

2. Beesley, *The Loss of the S.S.* Titanic, 53.

3. Beesley, *The Loss of the S.S.* Titanic, 58.

4. Beesley, *The Loss of the S.S.* Titanic, 61.

5. Beesley, *The Loss of the S.S.* Titanic, 62.

6. Beesley, *The Loss of the S.S.* Titanic, 34.

7. Jack Thayer, *The Sinking of SS* Titanic (Bath, UK: Spitfire, 2019), 18.

8. Thayer, *The Sinking of SS* Titanic, 19.

9. Thayer, *The Sinking of SS* Titanic, 19.

10. Thayer, *The Sinking of SS* Titanic, 19.

11. Thayer, *The Sinking of SS* Titanic, 19.

12. Thayer, *The Sinking of SS* Titanic, 19.

13. Thayer, *The Sinking of SS* Titanic, 20.

14. Beesley, *The Loss of the S.S.* Titanic, 36.

15. Daniel Butler, *Unsinkable: The Full Story of the RMS* Titanic (Mechanicsburg, PA: Stackpole Books, 1998), 83.

16. Butler, *Unsinkable*, 83.

17. Walter Lord, *A Night to Remember* (New York: Henry Holt and Company, 1955), 28.

18. Lord, *A Night to Remember*, 28.
19. Lord, *A Night to Remember*, 28.
20. Julie Lawson, *Ghosts of the* Titanic (Markham, ON: Scholastic Canada, 2011), 150.
21. Butler, *Unsinkable*, 84.
22. Lord, *A Night to Remember*, 28.
23. Tony Aspler, *Titanic* (New York: Doubleday, 1990), 296.
24. Lord, *A Night to Remember*, 29.
25. Lord, *A Night to Remember*, 32.

CHAPTER SIX

1. Daniel Gaetán-Beltrán, ed., *The* Titanic (Farmington Hills, MI: Greenhaven Press, 2015), 34.
2. Bonnie Buxton, *Damaged Angels* (Toronto: Knopf Canada, 2010), 39.
3. "Have Struck an Iceberg, Ship Is Listing Badly," *Bangor Daily News*, April 14, 1967.
4. "Have Struck an Iceberg."
5. John Eaton, Titanic*: Triumph and Tragedy* (New York: W. W. Norton, 1996), 166.

CHAPTER SEVEN

1. Charles Lightoller, Titanic *and Other Ships* (Oxford, UK: Oxford City Press, 2010), 174.
2. Lightoller, Titanic *and Other Ships*, 174.
3. John Welshman, Titanic*: The Last Night of a Small Town* (Oxford, UK: Oxford University Press, 2012), 113.
4. Jack Winocour, ed., *The Story of the* Titanic *as Told by Its Survivors* (New York: Dover Publications, 1960), 285.
5. Winocour, *The Story of the* Titanic, 285.
6. Winocour, *The Story of the* Titanic, 285.
7. Winocour, *The Story of the* Titanic, 286.
8. Walter Lord, *A Night to Remember* (New York: Henry Holt and Company, 1955), 36.
9. David Gleicher, *The Rescue of the Third Class on the* Titanic (Liverpool, UK: Liverpool University Press, 2017), 65.
10. Gleicher, *The Rescue of the Third Class*, 65.
11. Gleicher, *The Rescue of the Third Class*, 65.
12. Winocour, *The Story of the* Titanic, 289.
13. Winocour, *The Story of the* Titanic, 288.
14. Lord, *A Night to Remember*, 37.
15. Lord, *A Night to Remember*, 37.
16. Lord, *A Night to Remember*, 37.
17. Lord, *A Night to Remember*, 37.
18. "*Titanic* Survivor's Story in Conflict with Legend," *Simpson's Leader-Times* (Kittanning, PA), April 14, 1962.
19. "*Titanic* Survivor's Story in Conflict."
20. "*Titanic* Survivor's Story in Conflict."

21. "*Titanic* Survivor's Story in Conflict."

22. "*Titanic* Survivor's Story in Conflict."

23. "*Titanic* Survivor's Story in Conflict."

24. Peter Padfield, *The* Titanic *and the* Californian (New York: John Day Company, 1966), 64.

25. United States Congress, *"Titanic" Disaster: Hearings before a Subcommittee of the Committee on Commerce, United States Senate, Sixty-Second Congress, Second Session, Pursuant to S. Res. 283, Directing the Committee on Commerce to Investigate the Causes Leading to the Wreck of the White Star Liner* "Titanic" (Washington, DC: U.S. Government Printing Office, 1912), 389.

26. Lord, *A Night to Remember*, 37.

27. Gary Cooper, *The Man Who Sank the* Titanic? *The Life and Times of Captain Edward J. Smith* (Alsager, Staffordshire, UK: Witan Books, 1992), 121.

Chapter Eight

1. Erik Larson, *Thunderstruck* (New York: Crown, 2006), 153.

2. Larson, *Thunderstruck*, 151.

Chapter Nine

1. Gareth Russell, *The Ship of Dreams: The Sinking of the* Titanic (New York: Atria Books, 2019), 210.

2. George Behe, *On Board RMS* Titanic (Stroud, UK: History Press, 2012), 13.

3. John B. Thayer, *The Sinking of the S.S.* Titanic (Chicago: Academy Chicago, 2005), 338.

4. Thayer, *The Sinking of the S.S.* Titanic, 338.

5. Thayer, *The Sinking of the S.S.* Titanic, 339.

6. Thayer, *The Sinking of the S.S.* Titanic, 339.

7. Thayer, *The Sinking of the S.S.* Titanic, 339.

8. Russell, *The Ship of Dreams*, 212.

9. Daniel Butler, *The Other Side of the Night: The* Carpathia, *the* Californian, *and the Night the* Titanic *Was Lost* (New York: Casemate, 2009), 15.

10. Walter Lord, *A Night to Remember* (New York: Henry Holt and Company, 1955).

11. Jack Winocour, ed., *The Story of the* Titanic *as Told by Its Survivors* (New York: Dover Publications, 1960), 315.

12. Commonwealth Shipping Committee, *Volume 76* (London: H.M. Stationery Office, 1912), 65.

13. United States Congress, *"Titanic" Disaster: Hearings before a Subcommittee of the Committee on Commerce, United States Senate, Sixty-Second Congress, Second Session, Pursuant to S. Res. 283, Directing the Committee on Commerce to Investigate the Causes Leading to the Wreck of the White Star Liner* "Titanic" (Washington, DC: U.S. Government Printing Office, 1912), 82.

14. Lord, *A Night to Remember*, 45.

CHAPTER TEN

1. United States Congress, "Titanic" *Disaster: Hearings before a Subcommittee of the Committee on Commerce, United States Senate, Sixty-Second Congress, Second Session, Pursuant to S. Res. 283, Directing the Committee on Commerce to Investigate the Causes Leading to the Wreck of the White Star Liner "Titanic"* (Washington, DC: U.S. Government Printing Office, 1912), 783.

2. United States Congress, "Titanic" *Disaster,* 783.

3. Senan Molony, Titanic *Scandal: The Trial of the* Mount Temple (Stroud, UK: Amberly, 2009).

CHAPTER ELEVEN

1. Peter Padfield, *The* Titanic *and the* Californian (London: Hodder and Stoughton, 1965), 66.

2. Lawrence Beesley, *The Loss of the S.S.* Titanic (Boston: Houghton Mifflin, 1912), 79.

3. Beesley, *The Loss of the S.S.* Titanic, 79.

4. Beesley, *The Loss of the S.S.* Titanic, 81.

5. Beesley, *The Loss of the S.S.* Titanic, 83.

6. Beesley, *The Loss of the S.S.* Titanic, 85.

7. Beesley, *The Loss of the S.S.* Titanic, 85.

8. Daniel Butler, *Unsinkable: The Full Story of the RMS* Titanic (Mechanicsburg, PA: Stackpole Books, 1998), 121.

9. Tom Stacey, *The* Titanic (San Diego, CA: Lucent Books, 1989), 35.

10. Walter Lord, *A Night to Remember* (New York: Henry Holt and Company, 1955), 50.

11. Lord, *A Night to Remember,* 58.

12. Lord, *A Night to Remember,* 47.

13. Veronica Hinke, *The Last Night on the* Titanic*: Unsinkable Drinking, Dining, and Style* (Washington, DC: Regnery, 2019), xiv.

14. Lord, *A Night to Remember,* 47.

15. Lord, *A Night to Remember,* 48.

16. Lord, *A Night to Remember,* 48.

17. Lord, *A Night to Remember,* 48.

18. Lord, *A Night to Remember,* 48.

19. Lord, *A Night to Remember,* 49.

20. John Welshman, Titanic*: The Last Night of a Small Town* (Oxford, UK: Oxford University Press, 2012), 136.

21. Welshman, Titanic*: The Last Night,* 136.

22. Jack Winocour, ed., *The Story of the* Titanic *as Told by Its Survivors* (New York: Dover Publications, 1960), 291.

23. Winocour, *The Story of the* Titanic, 293.

24. Padfield, *The* Titanic *and the* Californian, 69.

25. "Woman Recalls Voyage of Unsinkable *Titanic,*" *Leader Times* (Kittanning, PA), April 14, 1962.

26. "Woman Recalls Voyage of Unsinkable *Titanic.*"

CHAPTER TWELVE

1. Daniel Butler, *The Other Side of the Night: The* Carpathia, *the* Californian, *and the Night the* Titanic *Was Lost* (New York: Casemate, 2009), 33.
2. Butler, *The Other Side of the Night*, 23.
3. Butler, *The Other Side of the Night*, 35.
4. Daniel Butler, *The Age of Cunard: A Transatlantic History* (Annapolis, MD: Lighthouse Press, 2004), 179.
5. Butler, *The Other Side of the Night*, 27.
6. Daniel Butler, *Unsinkable: The Full Story of RMS* Titanic (Mechanicsburg, PA: Stackpole Books, 1998), 114.
7. Butler, *Unsinkable*, 114.
8. Butler, *Unsinkable*, 114.
9. Great Britain Commissioner of Wrecks, *Formal Investigation into the Loss of the* Titanic (Chicago: University of Chicago Press, 1912), 697.
10. Butler, *The Other Side of the Night*, 64.
11. Butler, *The Other Side of the Night*, 64.
12. Daniel Gaetán-Beltrán, ed., *The* Titanic (Farmington Hills, MI: Greenhaven Press, 2015), 35.
13. Butler, *The Age of Cunard*, 180.
14. Butler, *The Age of Cunard*, 181.

CHAPTER THIRTEEN

1. "Local Wireless Operators Hear Disaster News," *Pittsburg Press*, April 17, 1912.
2. Commonwealth Shipping Committee, *Volume 76* (London: H.M. Stationery Office, 1912), 66.
3. Commonwealth Shipping Committee, *Volume 76*, 66.
4. Commonwealth Shipping Committee, *Volume 76*, 66.

CHAPTER FOURTEEN

1. United States Congress, "Titanic" *Disaster: Hearings before a Subcommittee of the Committee on Commerce, United States Senate, Sixty-Second Congress, Second Session, Pursuant to S. Res. 283, Directing the Committee on Commerce to Investigate the Causes Leading to the Wreck of the White Star Liner "Titanic"* (Washington, DC: U.S. Government Printing Office, 1912), 81.
2. Simon Angel, *The* Titanic: *"Everything Was against Us"* (N.p.: CreateSpace, 2012), 298.
3. Angel, *The* Titanic: *Everything*, 298.
4. John P. Eaton, Titanic: *Triumph and Tragedy* (New York: W. W. Norton, 1995), 166.
5. United States Congress, "Titanic" *Disaster*, 81.
6. United States Congress, "Titanic" *Disaster*, 929.

CHAPTER FIFTEEN

1. Lawrence Beesley, *The Loss of the S. S.* Titanic (Boston: Houghton Mifflin, 1912), 88.
2. Jack Winocour, ed., *The Story of the* Titanic *as Told by Its Survivors* (New York: Dover Publications, 1960), 38.

3. Winocour, *The Story of the* Titanic, 39.

4. Winocour, *The Story of the* Titanic, 39.

5. Winocour, *The Story of the* Titanic, 40.

6. Winocour, *The Story of the* Titanic, 40.

7. Winocour, *The Story of the* Titanic, 40.

8. Winocour, *The Story of the* Titanic, 40.

9. Winocour, *The Story of the* Titanic, 41.

10. Winocour, *The Story of the* Titanic, 41.

11. Beesley, *The Loss of the S.S.* Titanic, 99.

12. Daniel Butler, *The Other Side of the Night: The* Carpathia, *the* Californian, *and the Night the* Titanic *Was Lost* (New York: Casemate, 2009), 83.

13. Nick Barratt, *Lost Voices from the* Titanic (New York: St. Martin's, 2010), 243.

14. Butler, *The Other Side of the Night*, 84.

15. Butler, *The Other Side of the Night*, 87.

16. Butler, *The Other Side of the Night*, 88.

17. Commissioner of Wrecks, *Formal Investigation into the Loss of the S.S.* Titanic (London: H.M. Stationery Office, 1912), 169.

18. John Wilson Foster, *Titanic* (New York: Penguin, 1999), 181.

19. *Journal of Commerce Report of the* Titanic *Inquiry* (London: Offices of the *Journal of Commerce*, 1912), 64.

20. Stephanie Barczewski, Titanic: *100th Anniversary Edition: A Night Remembered* (New York: Continuum, 2011), 36.

21. Peter Padfield, *The* Titanic *and the* Californian (New York: John Day Company, 1966), 69.

22. Gary Cooper, *The Man Who Sank the* Titanic (Alsager, Staffordshire, UK: Witan Books, 1992), 112.

23. Cooper, *The Man Who Sank the* Titanic, 112.

24. Archibald Gracie, Titanic: *A Survivor's Story* (Chicago: Academy Chicago Publishers, 2005), 341.

25. Simon Angel, *The* Titanic: *"Everything Was against Us"* (N.p.: CreateSpace, 2012), 238.

26. Angel, *The* Titanic: *Everything*, 240.

27. Veronica Hinke, *The Last Night on the* Titanic (Washington, DC: Regnery, 2019), 123.

28. Hinke, *Last Night on the* Titanic, 121.

29. United States Congress, "Titanic" *Disaster: Hearings before a Subcommittee of the Committee on Commerce, United States Senate, Sixty-Second Congress, Second Session, Pursuant to S. Res. 283, Directing the Committee on Commerce to Investigate the Causes Leading to the Wreck of the White Star Liner* "Titanic" (Washington, DC: U.S. Government Printing Office, 1912), 772.

30. United States Congress, "Titanic" *Disaster*, 772.

31. John Welshman, Titanic: *The Last Night of a Small Town* (Oxford, UK: Oxford University Press, 2012), 135.

32. Walter Lord, *A Night to Remember* (New York: Henry and Holt Company, 1955), 54.

33. Lord, *A Night to Remember*, 54.

34. Lord, *A Night to Remember*, 54.
35. Lord, *A Night to Remember*, 54.
36. Lord, *A Night to Remember*, 54.
37. Gracie, Titanic: *A Survivor's Story*, 341.
38. Gracie, Titanic: *A Survivor's Story*, 341.

CHAPTER SIXTEEN

1. *Loss of the Steamship Titanic, Report of a Formal Investigation . . . as Conducted by the British Government* (Riverside, CT: 7 C's Press, 1912), 81.
2. Wade Sisson, *Racing through the Night:* Olympic's *Attempt to Reach* Titanic (Stroud, UK: Amberly, 2011), 11.
3. Sisson, *Racing through the Night.*
4. Sisson, *Racing through the Night.*
5. John P. Eaton, Titanic *Triumph and Tragedy* (New York: W. W. Norton, 1995), 166.
6. United States Congress, "Titanic" *Disaster: Hearings before a Subcommittee of the Committee on Commerce, United States Senate, Sixty-Second Congress, Second Session, Pursuant to S. Res. 283, Directing the Committee on Commerce to Investigate the Causes Leading to the Wreck of the White Star Liner* "Titanic" (Washington, DC: U.S. Government Printing Office, 1912), 739.
7. United States Congress, "Titanic" *Disaster*, 739.
8. United States Congress, "Titanic" *Disaster*, 1,135.

CHAPTER SEVENTEEN

1. Lawrence Beesley, *The Loss of the S.S.* Titanic (Boston: Houghton Mifflin, 1912), 37.
2. Beesley, *The Loss of the S.S.* Titanic, 37.
3. Beesley, *The Loss of the S.S.* Titanic, 46.
4. Walter Lord, *A Night to Remember* (New York: Henry Holt and Company, 1955), 54.
5. Lord, *A Night to Remember*, 56.
6. David Gleicher, *The Rescue of the Third Class on the* Titanic (Liverpool, UK: Liverpool University Press, 2017), 16.
7. Daniel Butler, *Unsinkable: The Full Story of the RMS* Titanic (Mechanicsburg, PA: Stackpole Books, 1998), 106.
8. Jack Winocour, ed., *The Story of the* Titanic *as Told by Its Survivors* (New York: Dover Publications, 1960), 294.
9. Charles Lightoller, Titanic *and Other Ships* (Oxford, UK: Oxford City Press, 2010), 185.
10. Lightoller, Titanic *and Other Ships*, 186.
11. Wyn Craig Wade, *The* Titanic: *End of a Dream* (New York: Rawson, Wade, 1979), 206.
12. United States Congress, "Titanic" *Disaster: Hearings before a Subcommittee of the Committee on Commerce, United States Senate, Sixty-Second Congress, Second Session, Pursuant to S. Res. 283, Directing the Committee on Commerce to Investigate the Causes Leading to the Wreck of the White Star Liner* "Titanic" (Washington, DC: U.S. Government Printing Office, 1912), 929.

13. Daniel Butler, *The Other Side of the Night: The* Carpathia, *the* Californian, *and the Night the* Titanic *Was Lost* (New York: Casemate, 2009), 73.

14. Butler, *The Other Side of the Night*, 74.

15. Butler, *The Other Side of the Night*, 74.

CHAPTER EIGHTEEN

1. "*Titanic* Call Ignored," *Evening Times-Republican* (Marshalltown, IA), April 26, 1912.

2. "Turned Away from the Sinking Ship," *Buffalo Commercial*, April 25, 1912.

3. "*Titanic* Call Ignored."

4. "*Titanic* Call Ignored."

5. "*Mount Temple* Passengers Saw Signals from *Titanic*," *Elmira* (New York) *Star Gazette*, April 25, 1912.

6. "Turned Away from Sinking Ship."

7. United States Congress, *"Titanic" Disaster: Hearings before a Subcommittee of the Committee on Commerce, United States Senate, Sixty-Second Congress, Second Session, Pursuant to S. Res. 283, Directing the Committee on Commerce to Investigate the Causes Leading to the Wreck of the White Star Liner "Titanic"* (Washington, DC: U.S. Government Printing Office, 1912), 1,098.

8. United States Congress, *"Titanic" Disaster*, 1,098.

9. Paul Lee, *The* Titanic *and the Indifferent Stranger* (Self-published, 2009), 38.

10. Lee, *The* Titanic *and the Indifferent Stranger*, 38.

11. Lee, *The* Titanic *and the Indifferent Stranger*, 38.

12. "Turned Away from Sinking Ship."

13. "Turned Away from Sinking Ship."

14. "Turned Away from Sinking Ship."

15. "Turned Away from Sinking Ship."

16. United States Congress, *"Titanic" Disaster*, 729.

17. "Turned Away from Sinking Ship."

18. United States Congress, *"Titanic" Disaster*, 772.

19. United States Congress, *"Titanic" Disaster*, 769.

20. Senan Molony, Titanic *Scandal: The Trial of the* Mount Temple (Stroud, UK: Amberly, 2009).

CHAPTER NINETEEN

1. Daniel Butler, *The Other Side of the Night: The* Carpathia, *the* Californian, *and the Night the* Titanic *Was Lost* (New York: Casemate, 2009), 76.

2. Butler, *The Other Side of the Night*, 77.

3. Butler, *The Other Side of the Night*, 78.

4. Butler, *The Other Side of the Night*, 78.

5. Butler, *The Other Side of the Night*, 78.

Chapter Twenty

1. John B. Thayer, Titanic: A Survivor's Story (Chicago: Academy Chicago Publishers, 2005), 342.

2. Thayer, Titanic: A Survivor's Story, 343.

3. Daniel Gaetán-Beltrán, ed., The Titanic (Farmington Hills, MI: Greenhaven Press, 2105), 141.

4. Jack Winocour, ed., The Story of the Titanic as Told by Its Survivors (New York: Dover Publications, 1960), 43.

5. Nick Barratt, Lost Voices from the Titanic (New York: St. Martin's, 2010), 106.

6. Lawrence Beesley, The Story of the S.S. Titanic (Boston: Houghton Mifflin, 1912), 44.

7. Archibald Gracie, The Truth about the Titanic (New York: Mitchell Kennerly, 1913), 82.

8. Beesley, The Story of the S.S. Titanic, 208.

9. Stephanie Barczewski, Titanic: 100th Anniversary Edition: A Night Remembered (New York: Continuum, 2011), 60.

10. Gracie, The Truth about the Titanic, 31.

11. United States Congress, "Titanic" Disaster: Hearings before a Subcommittee of the Committee on Commerce, United States Senate, Sixty-Second Congress, Second Session, Pursuant to S. Res. 283, Directing the Committee on Commerce to Investigate the Causes Leading to the Wreck of the White Star Liner "Titanic" (Washington, DC: U.S. Government Printing Office, 1912), 1,103.

12. United States Congress, "Titanic" Disaster, 1,103.

13. United States Congress, "Titanic" Disaster, 1,103.

14. Stephen Bottomore, The Titanic and Silent Cinema (Hastings, UK: Projection Box, 2000), 25.

15. Gaetán-Beltrán, ed., The Titanic, 130.

16. Donald Lynch, Titanic: An Illustrated History (New York: Hyperion, 1992), 134.

17. Lynch, Titanic: An Illustrated History, 134.

18. Thomas B. Costain and Douglas Beecroft, eds., 30 Stories to Remember (New York: Doubleday, 1962), 98.

19. United States Congress, "Titanic" Disaster, 591.

20. Daniel Butler, Unsinkable: The Full Story of the RMS Titanic (Mechanicsburg, PA: Stackpole Books, 1998), 121.

21. Gracie, The Truth about the Titanic, 168.

22. Costain and Beecroft, ed., 30 Stories to Remember, 98.

23. United States Congress, "Titanic" Disaster, 886.

24. Gracie, The Truth about the Titanic, 179.

25. Thayer, Titanic: A Survivor's Story, 343.

26. Thayer, Titanic: A Survivor's Story, 343.

27. Thayer, Titanic: A Survivor's Story, 343.

28. Thayer, Titanic: A Survivor's Story, 343.

29. Thayer, Titanic: A Survivor's Story, 343.

CHAPTER TWENTY-ONE

1. Charles Edward Crane, *Pendrift: Amenities of Column Conducting* (Brattleboro, VT: Stephen Daye Press, 1931).
2. United States Congress, *"Titanic" Disaster: Hearings before a Subcommittee of the Committee on Commerce, United States Senate, Sixty-Second Congress, Second Session, Pursuant to S. Res. 283, Directing the Committee on Commerce to Investigate the Causes Leading to the Wreck of the White Star Liner "Titanic"* (Washington, DC: U.S. Government Printing Office, 1912), 175.
3. Anton Gill, Titanic: *Building the World's Most Famous Ship* (New York: Transworld Publishers, 2013), 167.
4. Wade Sisson, *Racing through the Night* (Stroud, UK: Amberly, 2011).

CHAPTER TWENTY-TWO

1. *Amateur Radio*, vol. 45, nos. 1–6: 39.
2. United States Congress, *"Titanic" Disaster: Hearings before a Subcommittee of the Committee on Commerce, United States Senate, Sixty-Second Congress, Second Session, Pursuant to S. Res. 283, Directing the Committee on Commerce to Investigate the Causes Leading to the Wreck of the White Star Liner "Titanic"* (Washington, DC: U.S. Government Printing Office, 1912), 929.
3. United States Congress, *"Titanic" Disaster*, 929.
4. Henry Young, ed., *Popular Electricity in Plain English*, vol. 5 (June 1912), 215.
5. Young, *Popular Electricity*, 215.
6. John P. Eaton and Charles A. Haas, Titanic: *A Journey through Time* (New York: W. W. Norton, 1999), 66.

CHAPTER TWENTY-THREE

1. United States Congress, *"Titanic" Disaster: Hearings before a Subcommittee of the Committee on Commerce, United States Senate, Sixty-Second Congress, Second Session, Pursuant to S. Res. 283, Directing the Committee on Commerce to Investigate the Causes Leading to the Wreck of the White Star Liner "Titanic"* (Washington, DC: U.S. Government Printing Office, 1912), 78.
2. United States Congress, *"Titanic" Disaster*, 78.
3. United States Congress, *"Titanic" Disaster*, 78.
4. Logan Marshall, *On Board the* Titanic (Mineola, NY: Dover Publications, 2012), 252.
5. Jay Mowbray, *Sinking of the* Titanic (Princeton, NJ: National Publishing Company, 1912), 7.
6. United States Congress, *"Titanic" Disaster*, 12.
7. Daniel Butler, *The Other Side of the Night: The* Carpathia, *the* Californian, *and the Night the* Titanic *Was Lost* (New York: Casemate, 2009), 92.
8. Daniel Butler, *Unsinkable: The Full Story of the RMS* Titanic (Mechanicsburg, PA: Stackpole Books, 1998), 163.
9. *Shipping Casualties (Loss of the Steamship "Titanic"): Report of a Formal Investigation* . . . (London: H.M. Stationery Office, 1912), 59.

Chapter Twenty-Four

1. Parliament, *Report on the Loss of the* Titanic (London: Sutton, 1990), x.

2. Archibald Gracie, *The Truth about the* Titanic (New York: Mitchell Kennerly, 1913), 65.

3. Stephanie Barczewski, Titanic*: 100th Anniversary Edition: A Night Remembered* (New York: Continuum, 2011), 26.

4. United States Congress, *"*Titanic*" Disaster: Hearings before a Subcommittee of the Committee on Commerce, United States Senate, Sixty-Second Congress, Second Session, Pursuant to S. Res. 283, Directing the Committee on Commerce to Investigate the Causes Leading to the Wreck of the White Star Liner "*Titanic*"* (Washington, DC: U.S. Government Printing Office, 1912), 82.

5. United States Congress, *"*Titanic*" Disaster*, 82.

6. United States Congress, *"*Titanic*" Disaster*, 82.

7. Walter Lord, *A Night to Remember* (New York: Henry Holt and Company, 1955), 74.

8. Jack Thayer, Titanic*: A Survivor's Story* (Chicago: Academy Chicago Publishers, 2005), 344.

9. Lord, *A Night to Remember*, 75.

10. Lawrence Beesley, *The Loss of the S.S.* Titanic (Boston: Houghton Mifflin, 1912), 114.

11. Jack Winocour, ed., *The Story of the* Titanic *as Told by Its Survivors* (New York: Dover Publications, 1960), 298.

12. Winocour, *The Story of the* Titanic, 298.

13. Stephen Hines, Titanic*: One Newspaper, Seven Days, and the Truth That Shocked the World* (Naperville, IL: Sourcebooks, 2011), 157.

14. Michael Davie, *The* Titanic*: The Full Story of a Tragedy* (London: Bodley Head, 1986), 139.

15. Winocour, *The Story of the* Titanic, 317.

16. Wyn Craig Wade, *The* Titanic*: End of a Dream* (New York: Penguin, 1992), 224.

17. John Wilson Foster, *The Age of* Titanic (Dublin: Merlin Publishing, 2002), 63.

18. Hines, Titanic*: One Newspaper, Seven Days*, 158.

19. Foster, *Age of* Titanic, 101.

Chapter Twenty-Five

1. Daniel Butler, *The Other Side of the Night: The* Carpathia, *the* Californian, *and the Night the* Titanic *Was Lost* (New York: Casemate, 2009) 120.

2. Butler, *The Other Side of the Night*, 120.

3. Stephanie Sammartino McPherson, *Iceberg Right Ahead! The Tragedy of the* Titanic (Minneapolis, MN: Twenty-First Century Books, 2011), 55.

Chapter Twenty-Six

1. Lawrence Beesley, *The Loss of the S.S.* Titanic (Boston: Houghton Mifflin, 1912), 41.

2. Beesley, *The Loss of the S.S.* Titanic, 41.

3. Beesley, *The Loss of the S.S.* Titanic, 47.

4. Beesley, *The Loss of the S.S.* Titanic, 47.

5. Beesley, *The Loss of the S.S.* Titanic, 47.

6. Daniel Butler, *Unsinkable: The Full Story of the RMS* Titanic (Mechanicsburg, PA: Stackpole Books, 1998), 133.

7. Butler, *Unsinkable*, 118.

8. John Wilson Foster, *The Age of* Titanic: *Cross-Currents in Anglo-American Culture* (Dublin: Merlin Publishing, 2002), 131.

9. Jack Winocour, ed., *The Story of the* Titanic *as Told by Its Survivors* (New York: Dover Publications, 1960), 298.

10. *The Radio Telegrapher*, volume 29 (1912): 516.

11. Foster, *The Age of* Titanic, 97.

12. Foster, *The Age of* Titanic, 97.

13. Winocour, *Story of the* Titanic, 298.

14. Winocour, *Story of the* Titanic, 298.

15. Winocour, *Story of the* Titanic, 298.

16. Winocour, *Story of the* Titanic, 299.

17. Jack Thayer, Titanic: *A Survivor's Story* (Chicago: Academy Chicago Publishers, 2005), 345.

18. Gareth Russell, *The Ship of Dreams* (New York: Atria Books, 2019), 248.

19. Robert Ballard, *Robert Ballard's* Titanic (New York: Barnes & Noble, 2007), 29.

20. Thayer, Titanic: *A Survivor's Story*, 347.

21. Thayer, Titanic: *A Survivor's Story*, 347.

22. Butler, *Unsinkable*, 135.

23. Walter Lord, *A Night to Remember* (New York: Henry Holt and Company, 1955), 83.

24. Butler, *Unsinkable*, 138.

25. Jack Thayer, *The Sinking of the S.S.* Titanic, *April 14–15, 1912* (Bath, UK: Spitfire Publishers, 2019), 40.

26. Butler, *Unsinkable*, 138.

27. "*Titanic* Survivor's Story in Conflict with Legend," *Simpsons Leader Times* (Kittanning PA), April 14, 1962.

CHAPTER TWENTY-SEVEN

1. Daniel Butler, *The Other Side of the Night: The* Carpathia, *the* Californian, *and the Night the* Titanic *Was Lost* (New York: Casemate, 2009), 81.

2. United States Congress, "Titanic" *Disaster: Hearings before a Subcommittee of the Committee on Commerce, United States Senate, Sixty-Second Congress, Second Session, Pursuant to S. Res. 283, Directing the Committee on Commerce to Investigate the Causes Leading to the Wreck of the White Star Liner* "Titanic" (Washington, DC: U.S. Government Printing Office, 1912), 132.

3. *Shipping Casualties (Loss of the Steamship "Titanic"): Report of a Formal Investigation . . .* (London: H.M. Stationery Office, 1912), 82.

4. United States Congress, "Titanic" *Disaster*, 82.

5. John Maxtone-Graham, Titanic *Tragedy* (New York: W. W. Norton, 2012), 41.

6. United States Congress, "Titanic" *Disaster*, 1,129.

Chapter Twenty-Eight

1. Gareth Russell, *The Ship of Dreams* (New York: Atria Books, 2019), 258.

2. Walter Lord, *A Night to Remember* (New York: Henry Holt and Company, 1955), 96.

3. Lord, *A Night to Remember*, 96.

4. Archibald Gracie, *The Truth about the* Titanic (New York: Mitchell Kennerly, 1913), 96.

5. Jack Winocour, ed., *The Story of the* Titanic *as Told by Its Survivors* (New York: Dover Publications, 1960), 159.

6. Winocour, *Story of the* Titanic, 159.

7. Lord, *A Night to Remember*, 99.

8. Daniel Butler, *The Other Side of the Night: The* Carpathia, *the* Californian *and the* Night the Titanic *Was Lost* (New York: Casemate, 2009), 94.

9. Lord, *A Night to Remember*, 84.

10. Winocour, *Story of* Titanic, 301.

11. Winocour, *Story of* Titanic, 301.

12. John B. Thayer, Titanic: *A Survivor's Story* (Chicago: Academy Chicago Publishers, 2005), 348.

13. United States Congress, "Titanic" *Disaster: Hearings before a Subcommittee of the Committee on Commerce, United States Senate, Sixty-Second Congress, Second Session, Pursuant to S. Res. 283, Directing the Committee on Commerce to Investigate the Causes Leading to the Wreck of the White Star Liner* "Titanic" (Washington, DC: U.S. Government Printing Office, 1912), 1,105.

14. Lawrence Beesley, *The Loss of the S.S.* Titanic (Boston: Houghton Mifflin, 1912), 121.

15. United States Congress, "Titanic" *Disaster*, 282.

16. United States Congress, "Titanic" *Disaster*, 818.

17. United States Congress, "Titanic" Disaster, 786.

18. Lord, *A Night to Remember*, 102.

19. Lord, *A Night to Remember*, 103.

Chapter Twenty-Nine

1. Daniel Butler, *Unsinkable: The Full Story of the RMS* Titanic (Mechanicsburg, PA: Stackpole Books, 1998), 146.

2. United States Congress, "Titanic" *Disaster: Hearings before a Subcommittee of the Committee on Commerce, United States Senate, Sixty-Second Congress, Second Session, Pursuant to S. Res. 283, Directing the Committee on Commerce to Investigate the Causes Leading to the Wreck of the White Star Liner* "Titanic" (Washington, DC: U.S. Government Printing Office, 1912), 777.

3. United States Congress, "Titanic" *Disaster*, 929.

4. United States Congress, "Titanic" *Disaster*, 1,128.

5. Daniel Butler, *The Other Side of the Night: The* Carpathia, *the* Californian, *and the* Night the Titanic *Was Lost* (New York: Casemate, 2009), 95.

6. Butler, *The Other Side of the Night*, 95.

CHAPTER THIRTY

1. Lawrence Beesley, *The Loss of the S.S.* Titanic (Boston: Houghton Mifflin, 1912), 118.

2. Jack Thayer, *The Sinking of the S.S.* Titanic, *April 14–15, 1912* (Bath, UK: Spitfire Publishers, 2019), 41.

3. Thayer, *The Sinking of the S.S.* Titanic, 42.

4. Thayer, *The Sinking of the S.S.* Titanic, 42.

5. Thayer, *The Sinking of the S.S.* Titanic, 44.

6. Charles Lightoller, Titanic *and Other Ships* (Oxford, UK: Oxford City Press, 2010), 191.

7. Gareth Russell, *The Ship of Dreams* (New York: Atria Books, 2019), 258.

8. Russell, *The Ship of Dreams*, 257.

9. Russell, *The Ship of Dreams*, 257.

10. Judith Geller, Titanic: *Women and Children First* (New York: W. W. Norton, 1998), 65.

11. Geller, Titanic: *Women and Children First*, 65.

12. Russell, *The Ship of Dreams*, 259.

13. Russell, *The Ship of Dreams*, 259.

14. Walter Lord, *A Night to Remember* (New York: Henry Holt and Company, 1955), 104.

15. Beesley, *The Loss of the S.S.* Titanic, 125.

16. Lord, *A Night to Remember*, 106.

17. Archibald Gracie, *The Truth about the* Titanic (New York: Mitchell Kennerly, 1913), 143.

18. Commissioner of Wrecks, *Formal Investigation into the Loss of the S.S.* Titanic (London: H.M. Stationery Office, 1912), 263.

19. Commissioner of Wrecks, *Formal Investigation*, 263.

20. Russell, *The Ship of Dreams*, 259.

21. Russell, *The Ship of Dreams*, 259.

22. Lord, *A Night to Remember*, 109.

CHAPTER THIRTY-ONE

1. *Shipping Casualties (Loss of the Steamship "Titanic"): Report of a Formal Investigation* . . . (London: H.M. Stationery Office, 1912), 59.

2. Daniel Butler, *The Other Side of the Night: The* Carpathia, *the* Californian, *and the Night the* Titanic *Was Lost* (New York: Casemate, 2009), 95.

3. Butler, *The Other Side of the Night*, 95.

4. Butler, *The Other Side of the Night*, 96.

5. Butler, *The Other Side of the Night*, 96.

6. Butler, *The Other Side of the Night*, 96.

7. Butler, *The Other Side of the Night*, 96.

CHAPTER THIRTY-TWO

1. Charles Lightoller, Titanic *and Other Ships* (Oxford, UK: Oxford City Press, 2010), 193.

2. Jack Thayer, *The Sinking of the S.S.* Titanic, *April 14–15, 1912* (Bath, UK: Spitfire Publishers, 2019), 44.

3. Thayer, *The Sinking of the S.S.* Titanic, 44.

4. Thayer, *The Sinking of the S.S.* Titanic, 46.

5. Daniel Butler, *The Other Side of the Night: The* Carpathia, *the* Californian, *and the Night the* Titanic *Was Lost* (New York: Casemate, 2009), 104.

6. Butler, *The Other Side of the Night*, 104.

7. Walter Lord, *A Night to Remember* (New York: Henry Holt and Company, 1955), 114.

8. Butler, *The Other Side of the Night*, 104.

9. Thayer, *The Sinking of the S.S.* Titanic, 46.

CHAPTER THIRTY-THREE

1. Lawrence Beesley, *The Loss of the S.S.* Titanic (Boston: Houghton Mifflin, 1912), 131.

2. Beesley, *The Loss of the S.S.* Titanic, 131.

3. Beesley, *The Loss of the S.S.* Titanic, 132.

4. Daniel Butler, *The Other Side of the Night: The* Carpathia, *the* Californian, *and the Night the* Titanic *Was Lost* (New York: Casemate, 2009), 105.

5. Butler, *The Other Side of the Night*, 105.

6. Butler, *The Other Side of the Night*, 105.

7. United States Congress, *"Titanic" Disaster: Hearings before a Subcommittee of the Committee on Commerce, United States Senate, Sixty-Second Congress, Second Session, Pursuant to S. Res. 283, Directing the Committee on Commerce to Investigate the Causes Leading to the Wreck of the White Star Liner "Titanic"* (Washington, DC: U.S. Government Printing Office, 1912), 1,102.

8. Walter Lord, *A Night to Remember* (New York: Henry Holt and Company, 1955), 125.

9. Butler, *The Other Side of the Night*, 105.

10. Butler, *The Other Side of the Night*, 105.

11. Butler, *The Other Side of the Night*, 106.

12. Butler, *The Other Side of the Night*, 107.

13. Beesley, *The Loss of the S.S.* Titanic, 134.

14. United States Congress, *"Titanic" Disaster*, 394.

15. Jack Winocour, ed., *The Story of the* Titanic *as Told by Its Survivors* (New York: Dover Publications, 1960), 234.

16. Beesley, *The Loss of the S.S.* Titanic, 136.

17. Lord, *A Night to Remember*, 130.

18. Butler, *The Other Side of the Night*, 109.

19. Butler, *The Other Side of the Night*, 109.

20. Butler, *The Other Side of the Night*, 109.

CHAPTER THIRTY-FOUR

1. Daniel Butler, *The Other Side of the Night: The* Carpathia, *the* Californian, *and the Night the* Titanic *Was Lost* (New York: Casemate, 2009), 98.

2. Butler, *The Other Side of the Night*, 99.

3. Daniel Butler, *Unsinkable: The Full Story of the RMS* Titanic (Mechanicsburg, PA: Stackpole Books, 1998), 165.

CHAPTER THIRTY-FIVE

1. George Behe, *Voices from the* Carpathia: *Rescuing RMS* Titanic (Cheltenham, UK: History Press, 2015), 46.

2. "Local Wireless Operators Hear Disaster News," *Pittsburg Press*, April 17, 1912.

3. Ella Wheeler Wilcox, *The Worlds and I* (New York: George H. Doran, 1918), 219.

4. Mark Chirnside, *The Olympic Class Ships:* Olympic, Titanic, Britannic (Cheltenham, UK: History Press, 2004), 79.

5. Chirnside, *The Olympic Class Ships*, 78.

6. Wilcox, *The Worlds and I*, 219.

7. Wade Sisson, *Racing through the Night* (Stroud, UK: Amberly, 2011), 75.

8. Tom Kuntz, ed., *The* Titanic *Disaster Hearings* (New York: Pocket Books, 1998), 113.

9. United States Congress, "Titanic" *Disaster: Hearings before a Subcommittee of the Committee on Commerce, United States Senate, Sixty-Second Congress, Second Session, Pursuant to S. Res. 283, Directing the Committee on Commerce to Investigate the Causes Leading to the Wreck of the White Star Liner* "Titanic" (Washington, DC: U.S. Government Printing Office, 1912), 176.

10. United States Congress, "Titanic" *Disaster*, 1,135.

11. *New York Herald*, April 15, 1912.

12. *Evening Sun*, April 15, 1912.

13. *New York Times*, April 15, 1912.

14. United States Congress, "Titanic" *Disaster*, 1,136.

15. United States Congress, "Titanic" *Disaster*, 1,136.

16. John P. Eaton and Charles A. Haas, Titanic: *Triumph and Tragedy*, 2nd ed. (New York: W. W. Norton, 1998), 204.

17. United States Congress, "Titanic" *Disaster*, 1,136.

18. United States Congress, "Titanic" *Disaster*, 1,136.

19. United States Congress, "Titanic" *Disaster*, 178.

20. Sisson, *Racing through the Night*, 76.

21. Sisson, *Racing through the Night*, 76.

22. Sisson, *Racing through the Night*, 76.

CHAPTER THIRTY-SIX

1. United States Congress, "Titanic" *Disaster: Hearings before a Subcommittee of the Committee on Commerce, United States Senate, Sixty-Second Congress, Second Session, Pursuant to S. Res. 283, Directing the Committee on Commerce to Investigate the Causes Leading to the Wreck of the White Star Liner* "Titanic" (Washington, DC: U.S. Government Printing Office, 1912), 929.

2. United States Congress, "Titanic" *Disaster*, 929.

3. Lawrence Beesley, *The Loss of the S.S.* Titanic (Boston: Houghton Mifflin, 1912), 147.

4. Walter Lord, *A Night to Remember* (New York: Henry Holt and Company, 1955), 134.

5. Lord, *A Night to Remember*, 134.

6. Beesley, *The Loss of the S.S.* Titanic, 200.

7. Lord, *A Night to Remember*, 134.

8. Jack Winocour, ed., *The Story of the* Titanic *as Told by Its Survivors* (New York: Dover Publications, 1960), 56.

9. Charles Lightoller, Titanic *and Other Ships* (Oxford, UK: Oxford City Press, 2010), 193.

10. Winocour, *The Story of the* Titanic, 56.

11. John B. Thayer, Titanic*: A Survivor's Story* (Chicago: Academy Chicago Publishers, 2005), 354.

12. Daniel Butler, *The Other Side of the Night: The* Carpathia, *the* Californian, *and the Night the* Titanic *Was Lost* (New York: Casemate, 2009), 111.

13. Lord, *A Night to Remember*, 135.

14. Thayer, Titanic*: A Survivor's Story*, 355.

15. Beesley, *The Loss of the S.S.* Titanic, 201.

16. Archibald Gracie, *The Truth about the* Titanic (New York: Mitchell Kennerly, 1913), 111.

17. Beesley, *The Loss of the S.S.* Titanic, 204.

18. Beesley, *The Loss of the S.S.* Titanic, 204.

19. Beesley, *The Loss of the S.S.* Titanic, 136.

20. Thayer, Titanic*: A Survivor's Story*, 355.

21. Andrew Wilson, *Shadow of the* Titanic (New York: Atria Books, 2012), 190.

Chapter Thirty-Seven

1. United States Congress, *"*Titanic*" Disaster: Hearings before a Subcommittee of the Committee on Commerce, United States Senate, Sixty-Second Congress, Second Session, Pursuant to S. Res. 283, Directing the Committee on Commerce to Investigate the Causes Leading to the Wreck of the White Star Liner "Titanic"* (Washington, DC: U.S. Government Printing Office, 1912), 1,137.

2. United States Congress, *"*Titanic*" Disaster*, 1,137.

3. Wade Sisson, *Racing through the Night* (Stroud, UK: Amberly, 2011), 90.

4. United States Congress, *"*Titanic*" Disaster*, 1,137.

5. Sisson, *Racing through the Night*, 91.

6. John P. Eaton and Charles A. Haas, Titanic*: Triumph and Tragedy*, 2nd ed.(New York: W. W. Norton, 1998), 209.

7. Wyn Craig Wade, *The* Titanic*: Disaster of a Century* (New York: Skyhorse Publishing, 2012), 32.

8. Wade, *The* Titanic*: Disaster of a Century*, 32.

9. Wade, *The* Titanic*: Disaster of a Century*, 33.

10. Wade, *The* Titanic*: Disaster of a Century*, 33.

11. Wade, *The* Titanic*: Disaster of a Century*, 33.

12. Theodore Dreiser, *A Traveler at Forty* (New York: Century, 1913), 521.

CHAPTER THIRTY-EIGHT

1. Jack Winocour, ed., *The Story of* Titanic *as Told by Its Survivors* (New York: Dover Publications, 1960), 78.

2. Winocour, *The Story of the* Titanic, 78.

3. Jack Thayer, *The Sinking of the S.S.* Titanic, *April 14–15, 1912* (Bath, UK: Spitfire Publishers, 2019), 48.

4. Andrew Wilson, *Shadow of the* Titanic (New York: Atria Books, 2012), 59.

5. Lawrence Beesley, *The Loss of the S.S.* Titanic (Boston: Houghton Mifflin, 1912), 210.

6. Commonwealth Shipping Committee, *Volume 76* (London: H.M. Stationery Office, 1912), 83.

7. Commonwealth Shipping Committee, *Volume 76*, 83.

8. Commonwealth Shipping Committee, *Volume 76*, 83.

9. United States Congress, "Titanic" *Disaster: Hearings before a Subcommittee of the Committee on Commerce, United States Senate, Sixty-Second Congress, Second Session, Pursuant to S. Res. 283, Directing the Committee on Commerce to Investigate the Causes Leading to the Wreck of the White Star Liner* "Titanic" (Washington, DC: U.S. Government Printing Office, 1912), 691.

CHAPTER THIRTY-NINE

1. "Tried to Pick Up Wreck Message," *Leavenworth Times*, April 17, 1912.

2. "Tried to Pick Up Wreck Message."

3. "Tried to Pick Up Wreck Message."

4. Walter Lord, *A Night to Remember* (New York: Henry Holt and Company, 1955), 166.

5. Mac Smith, *Mainers on the* Titanic (Camden, ME: Down East Books, 2014), 17.

6. Lord, *A Night to Remember*, 141.

7. Lord, *A Night to Remember*, 141.

8. Lord, *A Night to Remember*, 141.

9. Lord, *A Night to Remember*, 141.

10. Wyn Craig Wade, *The* Titanic*: End of a Dream* (New York: Rawson, Wade, 1979), 43.

11. United States Congress, "Titanic" *Disaster: Hearings before a Subcommittee of the Committee on Commerce, United States Senate, Sixty-Second Congress, Second Session, Pursuant to S. Res. 283, Directing the Committee on Commerce to Investigate the Causes Leading to the Wreck of the White Star Liner* "Titanic" (Washington, DC: U.S. Government Printing Office, 1912), 478.

12. Wade, *The* Titanic*: End of Dream*, 46.

CHAPTER FORTY

1. Lawrence Beesley, *The Loss of the S.S.* Titanic (Boston: Houghton Mifflin, 1912), 77.

2. Walter Lord, *A Night to Remember* (New York: Henry Holt and Company, 1955), 146.

3. Daniel Butler, *The Other Side of the Night: The* Carpathia, *the* Californian, *and the Night the* Titanic *Was Lost* (New York: Casemate, 2009), 118.

4. United States Congress, "Titanic" *Disaster: Hearings before a Subcommittee of the Committee on Commerce, United States Senate, Sixty-Second Congress, Second Session, Pursuant to S. Res. 283, Directing the Committee on Commerce to Investigate the Causes Leading to the Wreck of the White Star Liner* "Titanic" (Washington, DC: U.S. Government Printing Office, 1912), 1,138.

5. United States Congress, "Titanic" *Disaster*, 1,138.

6. Beesley, *The Loss of the S.S.* Titanic, 77.

7. United States Congress, "Titanic" *Disaster*, 1,139.

8. Eric Clements, *Captain of the* Carpathia (New York: Bloomsbury, 2016), 62.

9. United States Senate Inquiry, Day 18, May 25, 1912, Titanic Inquiry Project, https://www.titanicinquiry.org/USInq/AmInq18header.php.

10. United States Congress, "Titanic" *Disaster*, 17.

11. "Canada Joins U.S. in *Titanic* Quiz," *Dixon* (Illinois) *Evening Telegraph*, April 26, 1912.

12. Stephen Hines, Titanic: *One Newspaper, Seven Days, and the Truth That Shocked the World* (Naperville, IL: Sourcebooks, 2011), 154.

13. "Canada Joins U.S. in *Titanic*."

14. United States Congress, "Titanic" *Disaster*, 846.

15. United States Congress, "Titanic" *Disaster*, 483.

16. Jack Winocour, ed., *The Story of the* Titanic *as Told by Its Survivors* (New York: Dover Publications, 1960), 313.

17. Grace Eckley, *Maiden Tribute: A Life of W. T. Stead* (Philadelphia, PA: Xlibris, 2007), 379.

18. John Booth and Sean Coughlan, Titanic: *Signals of Disaster* (N.p.: White Star Publications, 1993), 129–32.

19. Beesley, *The Loss of the S.S.* Titanic, 220.

20. Beesley, *The Loss of the S.S.* Titanic, 221.

21. Beesley, *The Loss of the S.S.* Titanic, 224.

22. Beesley, *The Loss of the S.S.* Titanic, 224.

23. Steve Turner, *The Band That Played On* (Nashville, TN: Thomas Nelson, 2012), 6.

Chapter Forty-One

1. Jack Winocour, ed., *The Story of the* Titanic *as Told by Its Survivors* (New York: Dover Publications, 1960), 314.

2. Winocour, *The Story of the* Titanic, 314.

3. Winocour, *The Story of the* Titanic, 317.

4. Winocour, *The Story of the* Titanic, 313.

5. Winocour, *The Story of the* Titanic, 313.

6. Winocour, *The Story of the* Titanic, 319.

7. Archibald Gracie, *The Truth about the* Titanic (New York: Mitchell Kennerly, 1913), 96.

8. Harold Bride, "Thrilling Story by *Titanic*'s Surviving Wireless Man," *New York Times*, April 19, 1912.

9. Bride, "Thrilling Story by *Titanic*'s Surviving Wireless Man."

10. "Wireless Operator Is Recovering," *Akron Beacon Journal*, April 20, 1912.

11. Paul Lee, *The* Titanic *and the Indifferent Stranger* (Self-published, 2009), 14.

12. *Clinton (Massachusetts) Daily*, April 23, 1912.

13. Daniel Butler, *The Other Side of the Night: The* Carpathia, *the* Californian, *and the Night the* Titanic *Was Lost* (New York: Casemate, 2009), 138.

14. Butler, *The Other Side of the Night*, 138.

15. Butler, *The Other Side of the Night*, 148.

16. United States Congress, *"Titanic" Disaster: Hearings before a Subcommittee of the Committee on Commerce, United States Senate, Sixty-Second Congress, Second Session, Pursuant to S. Res. 283, Directing the Committee on Commerce to Investigate the Causes Leading to the Wreck of the White Star Liner "Titanic"* (Washington, DC: U.S. Government Printing Office, 1912), 696.

17. United States Congress, *"Titanic" Disaster*, 696.

18. United States Congress, *"Titanic" Disaster*, 79.

19. United States Congress, *"Titanic" Disaster*, 11.

20. *Loss of the Steamship Titanic, Report of a Formal Investigation . . . as Conducted by the British Government* (Riverside, CT: 7 C's Press, 1912), 60.

21. United States Congress, *"Titanic" Disaster*, 783.

22. Daniel Butler, *Unsinkable: The Full Story of the RMS* Titanic (Mechanicsberg, PA: Stackpole Books, 1998), 188.

23. Butler, *Unsinkable*, 189.

24. Butler, *The Other Side of the Night*, 186.

25. Butler, *The Other Side of the Night*, 186.

26. Butler, *The Other Side of the Night*, 184.

Bibliography

"Aid Reaches *Titanic* Passengers Taken Off." *Brooklyn Daily Times*, April 15, 1912.

Allinson, Robert Elliott. *Saving Human Lives: Lessons in Management Ethics*. New York: Springer, 2005.

Amateur Radio, volume 45, numbers 1–6: 39.

Angel, Simon. *The* Titanic: *"Everything Was against Us."* N.p.: CreateSpace, 2012.

"Anxious Queries Reach *Carpathia*." *Pittsburgh Gazette Times*, April 18, 1912.

Aspler, Tony. *Titanic*. New York: Doubleday, 1990.

Ballard, Robert. *The Discovery of the Titanic*. New York: Warner, 1987.

———. *Robert Ballard's* Titanic: *Exploring the Greatest of All Lost Ships*. New York: Barnes & Noble, 2007.

Barczewski, Stephanie. Titanic: *100th Anniversary Edition: A Night Remembered*. New York: Continuum, 2011.

Barratt, Nick. *Lost Voices from the* Titanic. New York: St. Martin's, 2010.

Beesley, Lawrence. *The Loss of the S.S.* Titanic: *Its Story and Its Lessons*. Boston: Houghton Mifflin, 1912.

Behe, George. *A Death on the Titanic: The Loss of Major Archibald Butt*. Self-published, 2011.

———. *On Board RMS* Titanic. Stroud, UK: History Press, 2012.

———. *Voices from the* Carpathia: *Rescuing RMS* Titanic. Cheltenham, UK: History Press, 2015.

Beveridge, Bruce. Titanic: *The Ship Magnificent*, 5th ed., 2 vols. Stroud, UK: History Press, 2016.

Bigham, Randy. *Finding Dorothy: A Biography of Dorothy Gibson*. London: Lulu, 2012.

Booth, John, and Sean Coughlan. Titanic: *Signals of Disaster*. N.p.: White Star Publications, 1993.

Bottomore, Stephen. *The* Titanic *and Silent Cinema*. Hastings, UK: Projection Box, 2000.

Brewster, Hugh. *Gilded Lives Fatal Voyage: The* Titanic's *First-Class Passengers and Their World*. New York: Broadway Paperbacks, 2012.

Bride, Harold. "Thrilling Story by *Titanic's* Surviving Wireless Man." *New York Times*, April 19, 1912.

Brown, David. *The Last Log of the* Titanic. Camden, ME: International Marine Press, 2001.

Brown, Richard. *Voyage of the Iceberg: The Story of the Iceberg That Sank the* Titanic. Toronto, ON: James Lorimer, 2012.

Bryceson, Dave. *The* Titanic *Disaster: As Reported in the British National Press, April–July 1912.* New York: W. W. Norton, 1997.

Bullock, Shan F. *A* Titanic *Hero: Thomas Andrews, Shipbuilder.* Riverside, CT: 7 C's Press, 1973.

Butler, Daniel. *The Age of Cunard: A Transatlantic History.* Annapolis, MD: Lighthouse Press, 2004.

———. *The Other Side of the Night: The* Carpathia, *the* Californian, *and the Night the* Titanic *Was Lost.* New York: Casemate, 2009.

———. *Unsinkable: The Full Story of the RMS* Titanic. Mechanicsburg, PA: Stackpole Books, 1998.

Buxton, Bonnie. *Damaged Angels.* Toronto: Knopf Canada, 2010.

Cameron, Stephen. Titanic*: Belfast's Own.* Dublin: Wolfhound Press, 1998.

"Canada Joins U.S. in *Titanic* Quiz." *Dixon* (Illinois) *Evening Telegraph*, April 26, 1912.

Cannon, Kenneth L., II. "Isaac Russell's Remarkable Interview with Harold Bride, Sole Surviving Wireless Operator from the *Titanic.*" *Utah Historical Quarterly*, vol. 81, no. 4 (Fall 2013): 325–44.

"*Carpathia* Here Tonight with *Titanic*'s Survivors." *New York Times*, April 18, 1912.

Chirnside, Mark. *The Olympic Class Ships:* Olympic, Titanic, Britannic. Cheltenham, UK: History Press, 2004.

Clements, Eric. *Captain of the* Carpathia*: The Seafaring Life of* Titanic *Hero Sir Arthur Henry Rostron.* New York: Bloomsbury, 2016.

Commissioner of Wrecks. *Formal Investigation into the Loss of the S.S.* Titanic. London: H.M. Stationery Office, 1912.

Commonwealth Shipping Committee. *Volume 76.* London: H.M. Stationery Office, 1912.

"Complete Story of the *Titanic* Disaster." *Philadelphia Inquirer*, April 21, 1912.

Compton, Nic. Titanic *on Trial: The Night the* Titanic *Sank Told through the Testimonies of Her Passengers and Crew.* London: Bloomsbury, 2012.

Cooper, Gary. *The Man Who Sank the* Titanic? *The Life and Times of Captain Edward J. Smith.* Alsager, Staffordshire, UK: Witan Books, 1992.

Coughlan, Sean. Titanic*: The Final Messages from a Stricken Ship.* London: BBC News, 2012.

Davie, Michael. *The* Titanic*: The Full Story of a Tragedy.* London: Bodley Head, 1986.

Dougherty, Terri, Sean Stewart Price, and Sean McCullum. *Eyewitness to* Titanic*: From Building the Great Ship to the Search for Its Watery Grave.* North Mankato, MN: Capstone, 2015.

Dreiser, Theodore. *A Traveler at Forty.* New York: Century, 1913.

Eaton, John P. Titanic*: Triumph and Tragedy.* New York: W. W. Norton, 1995.

Eaton, John P., and Charles Haas. Titanic*: Destination Disaster.* New York: W. W. Norton, 1987.

———. Titanic*: A Journey through Time.* New York: W. W. Norton, 1999.

Eckley, Grace. *Maiden Tribute: A Life of W. T. Stead.* Philadelphia, PA: Xlibris, 2007.

"Feared Davits Would Break." *New York Sun*, May 10, 1912.

Foster, John Wilson. *The Age of* Titanic*: Cross-Currents in Anglo-American Culture*. Dublin: Merlin Publishing, 2002.

Gaetán-Beltrán, Daniel, ed. *The* Titanic. Farmington Hills, MI: Greenhaven Press, 2015.

Gardiner, Robin, and Dan van der Vat. *The* Titanic *Conspiracy: Cover-ups and Mysteries of the World's Most Famous Sea Disaster*. New York: Carol Publishing Group, 1995.

Geller, Judith. Titanic*: Women and Children First*. New York: W. W. Norton, 1998.

Gill, Anton. Titanic*: Building the World's Most Famous Ship*. New York: Transworld Publishers, 2013.

Gleicher, David. *The Rescue of the Third Class on the* Titanic. Liverpool, UK: Liverpool University Press, 2017.

Gracie, Archibald. Titanic*: A Survivor's Story*. Chicago: Academy Chicago Publishers, 2005.

———. *The Truth about the* Titanic. New York: Mitchell Kennerly, 1913.

Halpern, Samuel. "Navigational Inconsistencies of the SS *Californian*." http://www .titanicology.com/Californian/Navigational_Incosistencies.pdf.

"Have Struck an Iceberg, Ship Is Listing Badly." *Bangor Daily News*, April 14, 1967.

Hines, Richard Davenport. Titanic *Lives: Migrants and Millionaires, Conmen and Crew*. London: Harper, 2012.

Hines, Stephen. Titanic*: One Newspaper, Seven Days, and the Truth That Shocked the World*. Naperville, IL: Sourcebooks, 2011.

Hinke, Veronica. *The Last Night on the* Titanic*: Unsinkable Drinking, Dining, and Style*. Washington, DC: Regnery, 2019.

Hoffman, William, and Jack Grimm. *Beyond Reach: The Search for the* Titanic. New York: Beaufort Books, 1982.

"Ice Barriers, Held Back Rescue Ships." *Indianapolis News*, April 27, 1912.

Journal of Commerce Report of the Titanic *Inquiry*. London: Offices of the *Journal of Commerce*, 1912.

Kuntz, Tom. *The* Titanic *Disaster Hearings*. New York: Pocket Books, 1998.

Layton, J. Kent. *The Edwardian Superliners: A Trio of Trios*. Stroud, UK: Amberly, 2013.

Lee, Paul. *The* Titanic *and the Indifferent Stranger: The Complete Story of the* Titanic *and the* Californian. Self-published, 2009.

Lightoller, Charles. Titanic *and Other Ships*. Oxford, UK: Oxford City Press, 2010.

"Local Wireless Operators Hear Disaster News." *Pittsburg Press*, April 17, 1912.

"Log Tells Tragic Story." *Baltimore Sun*, April 26, 1912.

Lord, Walter. *The Night Lives On*. New York: William Morrow, 1986.

———. *A Night to Remember*. New York: Henry Holt and Company, 1955.

Loss of the Steamship Titanic, Report of a Formal Investigation . . . as Conducted by the British Government. Riverside, CT: 7 C's Press, 1912.

Lynch, Donald, and Ken Marschall (illustrator). Titanic*: An Illustrated History*. London: Hodder and Stoughton, 1992.

Marcus, Geoffrey. *The Maiden Voyage*. New York: Viking, 1969.

Marriott, John. *Disaster at Sea*. New York: Hippocrene Books, 1987.

Marriott, Leo. Titanic. London: Promotional Reprint Book Co., 1997.

Marshall, Logan. *On Board the* Titanic. Mineola, NY: Dover Publications, 2012.

Maxtone-Graham John. Titanic *Tragedy: A New Look at the Lost Liner*. New York: W. W. Norton, 2012.

Mayo, Jonathan. Titanic*: Minute by Minute*. London: Short Books, 2016.

McCluskie, Tom. *Anatomy of the* Titanic. London: PRC Publishing, 1998.

McPherson, Stephanie Sammartino. *Iceberg Right Ahead: The Tragedy of the* Titanic. Minneapolis, MN: Twenty-First Century Books, 2011.

Meredith, Lee William. *1912: Facts about the* Titanic. Mason City, IA: Savas Publishing, 1999.

Molony, Senan. Titanic *Scandal: The Trial of the* Mount Temple. Stroud, UK: Amberly, 2009.

"*Mount Temple* Passengers Saw Signals from *Titanic.*" *Elmira* (New York) *Star Gazette*, April 25, 1912.

Mowbray, Jay. *Sinking of the* Titanic. Princeton, NJ: National Publishing Company, 1912.

O'Connor, Richard. *Down to Eternity: How the Proper Edwardian and His World Died with the* Titanic. New York: William Morrow, 1957.

O'Donnell, E. E. *The Last Days of the* Titanic*: Photographs and Mementos of the Tragic Maiden Voyage*. Dublin: Wolfhound Press, 1997.

Oldham, Wilton J. *The Ismay Line*. Liverpool, UK: Journal of Commerce, 1961.

Padfield, Peter. *The* Titanic *and the* Californian. New York: John Day Company, 1966.

Pellegrino, Charles. *Her Name* Titanic*: The Untold Story of the Sinking and Finding of the Unsinkable Ship*. New York: Avon Books, 1988.

Quinn, Paul. Titanic *at Two A.M.: An Illustrated Narrative with Survivor Accounts*. Saco, ME: Fantail, 1997.

Raboy, Marc. *Marconi: The Man Who Networked the World*. Oxford, UK: Oxford University Press, 2016.

Radio Telegrapher, Volume 29, 1912.

Russell, Gareth. *The Ship of Dreams: The Sinking of the* Titanic *and the End of the Edwardian Era*. New York: Atria Books, 2019.

Shipping Casualties (Loss of the Steamship "Titanic"): Report of a Formal Investigation . . . London: H.M. Stationery Office, 1912.

"Ships Scouting for Possible Survivors." *Washington Herald*, April 16, 1912.

Simpson, Colin. *The* Lusitania*: Finally the Startling Truth about One of the Most Fateful of All Disasters of the Sea*. Boston: Little, Brown and Company, 1972.

Sisson, Wade. *Racing through the Night:* Olympic's *Attempt to Reach* Titanic. Stroud, UK: Amberly, 2011.

Stacey, Tom. *The* Titanic. San Diego, CA: Lucent Books, 1989.

"Steamer *Titanic* Rams Big Iceberg, Passengers Safe." *Moline* (Illinois) *Daily Dispatch*, April 15, 1912.

Stenson, Patrick. *The Odyssey of C. H. Lightoller*. New York: W. W. Norton, 1984.

Thayer, Jack. *The Sinking of the S.S.* Titanic, *April 14–15, 1912*. Bath, UK: Spitfire Publishers, 2019.

Thayer, John B. *The Sinking of the S.S.* Titanic, 2nd ed. Chicago: Chicago Review Press, 2005.

"Thrilling Story by *Titanic*'s Surviving Wireless Man." *New York Times*, April 19, 1912.

"*Titanic* Call Ignored." *Evening Times-Republican* (Marshalltown, IA), April 26, 1912.

Titanic Inquiry Project, U.S. Senate, Day 18, May 25, 1912. https://www.titanicinquiry .org/USInq/AmInq18header.php.

"*Titanic* Probe Will Result in New Laws." *Scranton Times*, April 27, 1912.

"*Titanic* Survivor's Story in Conflict with Legend." *Simpson's Leader-Times* (Kittanning, PA), April 14, 1962.

"Tried to Pick Up Wireless Message." *Leavenworth* (Kansas) *Times*, April 17, 1912.

"Turned Away from the Sinking Ship." *Buffalo* (New York) *Commercial*, April 25, 1912.

Turner, Steve. *The Band That Played On.* Nashville, TN: Thomas Nelson, 2012.

"Turning Dying Wails into Dollars and Cents." *Daily Gate City* (Keokuk, IA), April 25, 1912.

Underwood, Lamar. *The Greatest Disaster Stories Ever Told.* Guilford, CT: Globe Pequot, 2003.

United States Congress. "*Titanic" Disaster: Hearings before a Subcommittee of the Committee on Commerce, United States Senate, Sixty-Second Congress, Second Session, Pursuant to S. Res. 283, Directing the Committee on Commerce to Investigate the Causes Leading to the Wreck of the White Star Liner "Titanic."* Washington, DC: U.S. Government Printing Office, 1912.

Wade, Wyn Craig. *The* Titanic*: End of a Dream.* New York: Rawson, Wade, 1979.

Wells, Susan. Titanic*: Legacy of the World's Greatest Ocean Liner.* New York: Time Life, 1997.

Welshman, John. Titanic*: The Last Night of a Small Town.* Oxford, UK: Oxford University Press, 2012.

Wilcox, Ella Wheeler. *The Worlds and I.* New York: George H. Doran, 1918.

Wilson, Andrew. *Shadow of the* Titanic*: The Extraordinary Stories of Those Who Survived.* New York: Atria Books, 2012.

Winocour, Jack, ed. *The Story of the* Titanic *as Told by Its Survivors.* New York: Dover Publications, 1960.

"Woman Recalls Voyage of Unsinkable *Titanic*." *Leader Times* (Kittanning, PA), April 14, 1962.

Young, Henry, ed. *Popular Electricity in Plain English*, vol. 5 (June 1912).

Index